UNIPA Springer Series

The **UNIPA Springer Series** publishes single and co-authored thematic collected volumes, monographs, handbooks and advanced textbooks on specific issues of particular relevance in six core scientific areas. The issues may be interdisciplinary or within one specific area of interest. Manuscripts are invited for publication in the following fields of study:

1- Clinical Medicine;
2- Biomedical and Life Sciences;
3- Engineering and Physical Sciences;
4- Mathematics, Statistics and Computer Science;
5- Business, Economics and Law;
6- Human, Behavioral and Social Sciences.

Manuscripts submitted to the series are peer reviewed for scientific rigor followed by the usual Springer standards of editing, production, marketing and distribution. The series will allow authors to showcase their research within the context of a dynamic multidisciplinary platform. The series is open to academics from the University of Palermo but also from other universities around the world. Both scientific and teaching contributions are welcome in this series. The editorial products are addressed to researchers and students and will be published in the English language.

The volumes of the series are single-blind peer-reviewed.

Book proposals can be submitted to the *UNIPA Springer Series Technical Secretariat* at unipaspringer@unipa.it
At the following link, you can find some specific information about the submission procedure to the Editorial Board: https://www.unipa.it/strutture/springer/

More information about this series at https://link.springer.com/bookseries/13175

Giuseppe Sciumè · Eleonora Riva Sanseverino ·
Pierluigi Gallo
Editors

A Practical Guide to Trading and Tracing for the Energy Blockchain

UNIVERSITÀ
DEGLI STUDI
DI PALERMO

Springer

Editors
Giuseppe Sciumè
Department of Engineering
University of Palermo
Palermo, Italy

Eleonora Riva Sanseverino
Department of Engineering
University of Palermo
Palermo, Italy

Pierluigi Gallo
Department of Engineering
University of Palermo
Palermo, Italy

ISSN 2366-7516 ISSN 2366-7524 (electronic)
UNIPA Springer Series
ISBN 978-3-030-96606-5 ISBN 978-3-030-96607-2 (eBook)
https://doi.org/10.1007/978-3-030-96607-2

This Springer imprint is published by the registered company Springer Nature Switzerland AG
The registered company address is: Gewerbestrasse 11, 6330 Cham, Switzerland

Preface

The residential sector accounts for approximately 30% of the electrical energy consumed in developed countries. This demand is currently covered not only by fossil fuels but also by renewable energy sources. In addition, the "Clean Energy Package" presented by the European Commission has promoted since 2016 through specific directives the use of renewable energy and energy efficiency with the objective of a production share of renewable energy sources of 32% in the EU's energy mix by 2030 and the increase of energy efficiency to 32.5% by 2030 as compared to the 1990 levels. More recently, as part of the "European Green Deal," the European Commission proposed to raise the 2030 greenhouse gas emission reduction target, including emissions and removals, from 40% to at least 55% compared to 1990. More recently, the European Commission has launched the first part of its operational measures in the "Fit for 55%" package that will be supportive of the European climate policy framework and put member states on the right track for the ambitious reduction of 55% in carbon emissions by 2030 and net-zero emissions by 2050. An integrated planning of actions considering climate, land use, energy, transport and taxation will bring communities, regions and member states in line with the 2030 and 2050 targets already agreed in the European Climate Law. These directives encourage production from renewable sources, but at the same time, they require a reduction in fossil fuel production, while contributing to the transformation of the electricity system from centralized to distributed. The increase in generation from renewable sources guarantees a reduction of polluting emissions, but due to its intermittent nature, such generation is difficult to manage and predict. This European framework requires the development of new energy policies at the national level able to reduce overall consumption and support the implementation of customer-centric control and management systems. It is thus necessary to study the feasibility of the different strategies available, especially for the involvement of residential users in services that can contribute to the management of the power grid. The advent of the blockchain technology has allowed the development of new business models for the electricity market, opening this world also to end users who previously did not have the opportunity to participate in energy trading and allowing the implementation of services to ensure the correct operation of the power grid. Thanks to its characteristics and especially thanks to the feature of being distributed over the grid, the blockchain could be the solution to the balancing problems caused by the penetration of unpredictable renewable sources and could make a significant contribution to achieving the 2030 and ultimately the 2050 targets. Out of all the applications in the energy sector, the most popular and developed so far are Peer-to-Peer, Vehicle-to-Grid and Demand Response applications. Peer-to-Peer energy trading concerns the direct energy exchange between end users, Vehicle-to-Grid applications concern the exchange of energy between electric vehicles and power grid for congestion resolution and ancillary services provision, while Demand Response is related to the possibility to modify the demand from end users. Demand Response does not concern energy exchanges but, such as Vehicle-to-Grid, the solution of congestion problems and the provision of fast response ancillary services. Demand Response allows consumers to respond to market signals by increasing or reducing their energy consumption, with the aim of responding to peaks in electricity supply or demand, contributing to greater network flexibility and stability and more efficient use of infrastructures and energy resources. This service represents an important resource to achieve the objectives to be reached by 2030, as reducing peaks in demand avoids the use of peak generators with the consequent reduction of emissions and energy prices. The use of the blockchain for Demand Response service certification allows to create a distributed system in which even residential customers can communicate with the system operator to provide their flexibility, in a secure, transparent and traceable way. This book addresses the development of a new methodology for the implementation of Demand Response and Vehicle-to-Grid programs using blockchain technology for their tracking and certification. A Smart Contract was designed and written to run Demand Response events, calculate the customer baseline, calculate the support provided by each customer toward the implementation of the required load curve change and remunerate each customer with utility tokens in proportion to its contribution. To test the Demand Response methodology, a Hyperledger Fabric network and a Smart Contract were implemented on four nodes of the Microgrid Laboratory of the Department of Energy Technology, University of Aalborg (DK). Subsequently, a realistic scenario including two consumption nodes was developed, using electronic power converters for household profile generation and Smart Meters for consumption profile measurement. Finally, further experiments are described in which also the use on an energy management system is integrated and accounts for the optimal utilization of energy resources in a residential house. The theoretical and experimental results show the feasibility of Distributed Ledger Technologies in the management of smart grids with minimal investment in new hardware, while allowing the active participation of residential customers in Demand Response programs in a more transparent and fair way. The book provides a research path in the area of the blockchain for the energy sector according to the following structure. Chapter 1 introduces the mean features, the application field and the fundamental principles of the blockchain technology. In Chap. 2, there is an overview of the Italian electricity market, the involved actors and all the necessary background for better understanding the following chapters. Also, it is explained how blockchain technology can contribute to the electricity market, the new business models, the energy regulations as well as the newly enabled energy scenarios. Chapter 3 describes in details the three most successful blockchain applications in the energy sector. Then, in Chap. 4, it is presented a blockchain architecture and its network for the execution of Demand Response events, without the need for intermediaries acting between the system operator and the end users. In Chap. 5, the experimental set-up used for testing a Demand Response event, the implementation of the blockchain network and the obtained results are described. Finally, in Chap. 6, it is presented and described the BloRin project, which aims to realize a blockchain-based technology platform for renewable energy deployment by implementing Demand Response and Vehicle-to-Grid services supported by a permissioned blockchain platform.

Palermo, Italy

Giuseppe Sciumè
Eleonora Riva Sanseverino
Pierluigi Gallo

Contents

Chapter 1
Introduction

G. Sciumè, E. Riva Sanseverino, P. Gallo, M. L. Di Silvestre, G. Zizzo, R. Musca, J. C. Vasquez, and J. M. Guerrero

Abstract In this chapter, the basic elements for framing the blockchain technology in the different application fields is given. The fundamental principles of its distributed operation are given against traditional centralized systems are described together with all the main components of its successful architectures. These components also physically give the possibility to implement the main features of the blockchain technology, ranging from the smart contracts to the consensus mechanisms. Finally, a general overview of the blockchain technology features against the European General Data Protection Regulation is given.

1.1 The Blockchain Technology

The blockchain techonology was introduced in 2008 with the Bitcoin, the first and most known application for crypto-currencies transaction. Along the years, many companies have been striving for adopting this technology in many fields such as: finance, e-voting, copyright, patent management, smart contracts, supply chain and energy market. Nevertheless, the blockchain has been associated only to the Bitcoin. Only recently, thanks to the progress achieved in different relevant areas it is recognized as a Distributed Ledger Technology (DLT), and thus characterized by a distributed register. Bitcoin is the first crypto-currency in the world. It was founded with the purpose to create an electronic payment system allowing the exchange of value without the need for a trusted third party. The conventional and common e-commerce system relies on financial institutions, which are the third parties needed for processing electronic payments. This latter system is centralized and suffers from the inherent weaknesses of a model based on the trust of one or more users to a single trusted third party. The direct consequence of this model is an increase in the mediation cost, which consequently increases the transaction cost, thus lim-

G. Sciumè (✉) · E. Riva Sanseverino · P. Gallo · M. L. Di Silvestre · G. Zizzo · R. Musca
University of Palermo, Viale delle Scienze, Building 9, 90128 Palermo, Italy
e-mail: giuseppe.sciume01@unipa.it

J. C. Vasquez · J. M. Guerrero
University of Aalborg, Pontoppidanstræde 111, 9220 Aalborg, Denmark

G. Sciumè et al. (eds.), *A Practical Guide to Trading and Tracing for the Energy Blockchain*, UNIPA Springer Series, https://doi.org/10.1007/978-3-030-96607-2_1

iting the possibility to implement small transactions. To address these issues, a new cutting-edge technology (the blockchain), based on cryptographic evidence instead of trusting a third party, was proposed allowing the exchange of crypto-currency (in the case of Bitcoin) between two unknown users. The blockchain technology allows events to be recorded sequentially along time and guarantees the immutability of the recorded events as each change requires resources and/or the approval of most of the users of the system and not of a single authority, as in the typical case of centralized systems. Such a system is thus useful when two people want to make a deal, but do not trust each other, as there is no need to rely on third parties to ensure user reliability [1]. In centralized systems, transactions of any asset are recorded in a centralized database managed by an entity, which theoretically has the authority to perform any operation on the data, and therefore must be trusted to avoid malicious actions. In the financial sector, this entity is represented by the bank or a credit agency, while in the electricity field the central role can be played by different entities depending on the transaction typology and on the market phase. As an example, the GME in Italy and the CRE in France manage financial and energy transactions in the wholesale energy market. The Transmission System Operator (Terna in Italy and RTE in France) on the other hand, verifies the technical feasibility of the financial and energy transactions. If congestions or other technical issues arise, these are solved in the balancing market. Besides, recent indications from the European Network of Transmission System Operators (ENTSO-E), are underlining the increasingly important role of Distribution System Operators (DSOs) to manage the local grid-connected energy resources in a flexible way in a wider flexibility market [2]. Transactions related to the distribution network (either financial or power transactions) are also recorded in centralized databases. The innovation introduced by the blockchain technology is data distribution and trust among the users network, overcoming the problems that afflict centralized systems. In distributed systems, each user or node has the same capacity and permission for access to the full information, thus providing a solid foundation for trust and making the system more democratic. Decentralized systems like blockchain are generally more secure because data is stored on all nodes composing the network. If a cyber-attack occurs, there is no single point of failure, so the network can continue to operate. Conversely, in a traditional centralized system, it is required to rely on a single central authority to control and manage user permissions in the system. It is easy to understand that this kind of systems is more exposed in the event of a cyber-attack, but is efficient when the party controlling the network is trustworthy and behaves in a honest manner.

Another advantage arising from the use of the blockchain technology is the standardization of the communication protocol in Business to Business (B2B) and Business to Consumer (B2C) transactions. When a multi-party environment is in place, such as it happens in power and energy systems applications (trading of energy is one case), there is a large number of interactions, which are typically operated with diverse communication protocols and technologies. Using the same communication technology would largely limit the integration costs both in B2B and B2C interactions. The case of electricity trading is exemplary, as in this case B2B interactions occur in the day-ahead energy market. In this case, the communication protocols and technologies typically used are indeed numerous. In the same way, many are the platforms for trading. The EPEX SPOT platform matches orders transparently to all parties. The EPEX SPOT platform is one of the many communication technologies involved as many others are utilized to transmit similar information to the European Commodity Clearing (ECC). Still other technologies are used by the ECC to get commitments from both the buyers and the sellers. The ECC also transfers and receives data from the TSO and this may happen by means of further communication technologies [3]. The blockchain technology adoption creates a domain showing the three main features of B2B integration in equilibrium (data format, business process and communication protocol). In this way, the communication protocol takes the right relevance and is homogeneous for all actors. The blockchain is classified as a DLT, in which the ledger is organized into blocks, that are shared and managed by the nodes in the network. The blockchain can be public or private depending on access to the ledger, while it is called *permissioned* or *permissionless* depending on who is authorized to have access and maintain the ledger. In few words, a blockchain is a chain of blocks and each block accomodates a number of transactions. At the same time, the whole information saved in the blockchain is shared among all users taking part to it. Each user has visibility on all transactions, in this way, it is possible to trace the history of the information. As a result, users are entirely in charge of their actions and contracts among them. This is true whatever is the role of participants: end users, businesses, certification authorities, operators, ... Each block is linked to the prior one through cryptographic functions named "Hash functions". Such functions let a message of any length, in this case data in a block, to be turned into a message of fixed length which is a digest called *hash*. The latter uniquely identifies the original message. Moreover, it is not easy to find two different messages that are turned into the same hash. This feature of cryptographic Hash functions is named 'collision resistance'. Once the hash function is applied to a given message, it is no more possible to return to the source data. The distinctive feature of these functions is that they can be easily calculated and always output the same result starting from the same input data (deterministic functions). Blockchain technology takes advantage of these functions to create the associated distributed ledger, as on each block header, a *hash* is calculated from the transaction data saved into the block itself plus the *hash* of the preceding block. As a result, each block saves data about the preceding block, namely the *hash*. And just like a chain, thanks to the hash functions, it contains information of all the previous blocks and transactions. The described chaining system makes the ledger unchangeable, all blocks in the network are copied, see Fig. 1.1.

The blockchain exceeds the peculiarities of a distributed database, i.e. a database not physically saved on a single machine (server). It is implemented on many machines (or physical storage units, i.e.: in a cloud system) at the same time, all the machines being synchronized on the same documents. In fact, unlike a database, blockchain transactions need to be validated before being part of new blocks, needing the approval from most of the network nodes. Once this approval arrives from the majority of nodes, it is hard to modify as it would require an incredibly large computational effort (by duplicating the whole upstream ledger) or again attaining the approval of most of the network nodes.

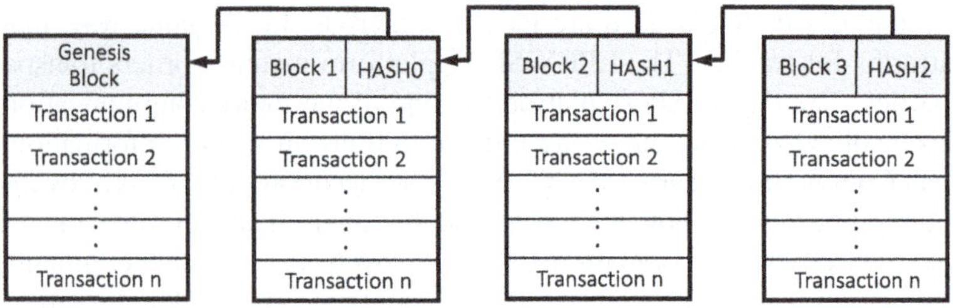

Fig. 1.1 Implementation of BC blocks

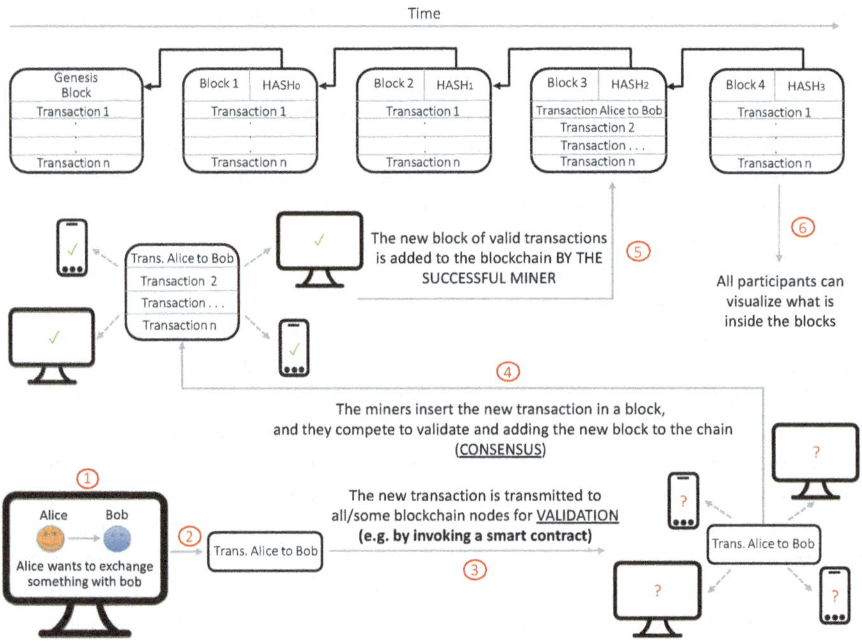

Fig. 1.2 How a blockchain with validation and consensus works

Figure 1.2 depicts an example of the process required to insert a new transaction in the blockchain. The procedure can be divided into six steps:

1. Request of transaction.
2. The new request of the transaction is sent to the network.
3. The transaction is sent to the nodes that are responsible for transactions—validation.
4. If the transaction is valid, the validator node inserts the new transaction in a new block containing other valid transactions and competes to validate the whole block (consensus process).

5. After the validation of the new block, the successful validator/miner can insert the new block in the chain.
6. Once the new block is added to the chain, all participants will be able to see its content.

Due to decentralization, it is also possible to prevent market abuse through monopolies and to reduce costs and regulatory oversight. The nodes that can issue the validation of single transactions and blocks (sets of transactions) are named "miners". The validation process can be implemented by means of particular algorithms called "Consensus Algorithms" or by running some pieces of code shared on the blockchain and named "Smart Contracts". Technicalities on how smart contracts work, on the different consensus algorithms and about the process for gaining consensus over a block of transactions are reported, with related references, in later sections. The so-called authoritative validation stems in a centralised system from a the validation delivered by a trusted authority or more in general by the so-called 'system manager'. The blockchain technology has the following features:

1. *Transparency* (in public blockchain), as each block that is added to the blockchain can be accessed by all participants and is in the archive of all participants.
2. *Trust*, based on cryptographic functions, which let the blockchain work in distributed and unreliable environments.
3. *Immutability*, since a block, thanks to the use of hash functions, once added, can be changed only having an extremely high computing power in case of permissionless blockchain (more than the one used for mining the single block) or with the approval of most of the participants in the network in the case of permissioned blockchain.
4. *Confidence*, as there is a shared reading among all the participants.
5. *Efficiency* (to be evaluated case by case), as it does not require intermediaries other than the classical transaction management system, thus simplifying processes, infrastructures and increasing operational efficiency.
6. *Control* and *Security*, as it imples the use of cryptography thus gaining greater data protection and lower risk of fraud.

As already said, the blockchain is not the only technology for storing data referring to transactions and using them. The papers in [4] and [5] offer a useful flow diagram for the assessment of whether the considered application needs a blockchain-based approach or not. In the flowchart, a few peculiarities are considered "basic requirements" for applying blockchain. These are:

- multiparty;
- decentralised operation;
- need for transactions history;
- no need for high speed for the process.

The blockchain is a time-stamped and immutable ledger that is distributed and managed by clusters of computers, as such it is a digital standard to ensure the security and immutability of shared data. Thanks to the benefits it provides compared to classic centralised systems, recently many companies are investing on the use of this new technology to develop and test many applications in different fields.

1.2 Application Fields

The word "blockchain" is synonymous of technological innovation, especially when used in those sectors where security of exchanged data, transparency and trust take priority. Blockchain is considered a next generation information technology with many potential benefits in several fields beyond digital currencies. As mentioned before, the first application of blockchain technology was the cryptocurrency Bitcoin, but blockchain represents a much bigger opportunity than Bitcoin. The most important feature of this new technology is the possibility to perform transactional activities of any kind without a trusted third party. So Bitcoin, and the technology introduced with it, is much more than a digital currency; the blockchain is a information technology. This technology is one of the first major identifiable implementations of decentralisation models that have the potential to reorganise all sorts of human activities thanks to their ability to provide frictionless and trust free interaction between people and technology [6, 7]. Below are the different sectors in which the use of blockchain technology is having the most success.

1.2.1 Cryptocurrency and Finance

The most widely accepted application for blockchain technology is in the field of finance, especially in the field of cryptocurrency, as it ensures the much appreciated transparency between the parties. Unlike classical currencies, cryptocurrencies do not involve coins or cash, as all payment units are exclusively digital. These cryptocurrencies are generally generated collectively across the whole system and, in most cases, when a cryptocurrency is launched, a defined number of units is established. The concept of "mining" (extraction) for the process of generating units has become widespread, and this explains why we often hear talk of "cryptocurrency extraction". Both the generation system and the individual operations are performed collectively using the blockchain of that cryptocurrency. The similarity with real currencies that include banknotes and coins is the fact that several cryptocurrencies, in addition to Bitcoin, such as Ethereum, have a countervalue (demand and users). If the units of a cryptocurrency possess a certain countervalue, convertible into central bank currencies such as Euro, the prerequisite for its use as a means of payment is basically fulfilled. But in order to actually pay with digital money, the seller must also accept the respective cryptocurrency as a means of payment. A pair of keys, one public and one private, is required to execute the payment. The public key is visible to everyone and has the same function as a bank account number: it serves as the starting address from which a user initiates a payment with the respective cryptocurrency. The private key is employed to recognize the transaction. Only the owner of the respective account can see the key used to sign the transaction. When making the payment, users only have to enter the amount and the destination address, i.e. the beneficiary's public key. To date, there are almost 3000 different cryptocurrencies, but not all of them are reliable or well established in the market. Among the most famous and used are Ethereum, Ripple, IOTA, Litecoin, Bitcoin Cash and Monero. The use of cryptocurrency allows you to take advantage of all the features provided by the blockchain, but there are also disadvantages such as large fluctuations in value, possible hacker attacks, exclusively virtual money, no access to the account after the loss of the private key and no loss insurance.

Else than its use in the area of cryptocurrency, the blockchain is increasingly being used in other financial services, such as stock exchanges, payments, repurchase agreements and digital identities. Many banks and financial institutions have also started to invest in blockchain in recent years [7–9].

1.2.2 Healthcare

The healthcare sector is also investing in blockchain projects. All the features of blockchain technology make it possible to create collaborative processes between multiple parties without the need for a third party and without necessarily trusting the players in the network, while at the same time ensuring transaction validity, tamper resistance, certainty of immutability, authenticity and origin of data and their integrity, transparency and ease of control. These features are very interesting for healthcare organizations. Thanks to the blockchain, the management of patients, medical records and health data promises to be more secure, as part of a sharable and unchangeable database that can be consulted by medical staff, while respecting patient privacy. In addition, medical records that are scattered due to transfer between different healthcare institutions can be combined to provide an opportunity to monitor the personal health of patients. It could facilitate the adoption of the Electronic Health File because, to date, each clinic/doctor's office generates a different file for the same patient. Thanks to the blockchain, it should be easier to archive the complete clinical history of patients, both those coming from health care facilities and from the Apps used by individuals to monitor pressure, heart rate, medication intake, etc. In this way the patient is also safer. His clinical history is shared and not modifiable, the doctors who treat him can verify, for example, which drugs have been prescribed, avoiding the use of incompatible drugs that are dangerous for his health. In addition, the blockchain could also offer benefits for health service payment processes, traceability of drug deliveries, validation of patient identity and consent to the processing of health data. Rather than having individual facilities manage the data, it would be possible to electronically attach consent to the data, thus giving the patient the opportunity to express consent for specific data analysis and processing purposes and specific privacy preferences. In Europe, there are many initiatives in this area [10, 11].

1.2.3 Agribusiness

It has always been very complex in the agri-food sector to maintain the authenticity and ownership of products, with the consequent introduction into the market of falsified products, uncertain prevention, imitations or unauthorized resale and processes that have not always guaranteed the complete transparency of the supply chain. Traceability of the field of 'food' is the ability to keep up with any food, feed or substance that is adequate for human consumption, as well as all the processes they have undergone, across all steps of the supply chain. Tracking food is for the benefit of both the consumer and the producer. The consumer, today more and more careful about the information about a certain product, wants to know in a clear and certified way information about the product he is consuming, while the producer would like to better manage stocks, reduce waste and open new market opportunities.

The blockchain system can be used as an innovative solution for the transparent certification of agricultural products, able to enhance the relationship of trust between the consumer and the agri-food chain, promoting high quality agricultural work as a guarantee of the food safety of the citizen and contributing to improve transparency and protection of the mark of origin for the benefit of consumers and businesses. In the agri-food sector, the blockchain makes it possible to trace a food product right from the cultivation and then from the geo-location of the cultivated land, retracing the supply chain to the retailer of this product in a meticulous and immutable way. In this way, an increase in security, speed and confidentiality of information throughout the supply can be achieved.

Since 2017, there has been a significant increase in blockchain applications in the agri-food sector [12, 13], mostly to support the traceability process. One of the most recent is SeedsBit [14], an innovative blockchain platform developed by the SNAPPLab of the University of Palermo for tracking and tracing of food products and related production processes.

The stakeholders have strong confidence in the guarantee of the immutability of the information promised by the blockchain platforms, in the improvement of the transparency along the supply chain and in increasing the effectiveness and efficiency of data recovery processes in case of food safety critical situations.

1.2.4 Insurance

Insurance policies are regulated by traditional contracts on paper, as a consequence, claims and payments are sometimes erroneous and often require human supervision. Moreover, most of the time, claims management is long and therefore expensive. The digitization of data and the use of blockchain in the insurance universe has the potential to automate many processes, such as claims management, thereby reducing the associated costs compared to traditional management. Blockchain technology and smart contracts can ensure that information is kept confidential and contracts are executed automatically in case that a claim occurs, allowing the optimization of issuing and managing policies, with an immediate and certain settlement of damage to policyholders. Smart contracts play a key role in this area, as they could for example encode the conditions to allow the transfer of the refund from the insurance company to the insured or activate an automatic refund transfer only if the customer, following a car crash, repairs the car at a car repair shop whose identity has been certified. The use cases are many and range from the processing of claims to the simplification of routine interactions and from risk prevention to property and accident insurance [15]. The use of the blockchain in the insurance industry would eliminate fraud and simplify many processes that are usually long and difficult to manage.

From a system architecture point of view, the most appropriate choice could be to adopt a combination of public and private blockchain, where the private blockchain, managed by the nodes of a trusted third party company, could be used to record claims policies and data, while the public blockchain could be used for reimbursement in terms of cryptocurrency tradable on the markets (e.g., Ether, Bitcoin). As an alternative, the insurance agency may take the decision to use only a public blockchain, agreeing to sustain more costly transaction costs so as to improve its standing and get customer confidence with the guarantee of full decentralisation.

Today, several companies, such as IBM, Allianz and KPMG are developing several projects in this area.

1.2.5 Copyright Protection

Property monitoring has always been a difficult topic, especially with the development of the internet and social media. It is well known that copyright is not always respected in many web applications, ranging from peer-to-peer file sharing services such as Napster to web-based photographs. From the point of view of a file holder, copyright is often ignored. As a result, unauthorized, and even illegal, file sharing and the use of copyrighted content is still a notable problem. Blockchain technology is a solution that can be applied to solve these problems. Taking into account that a file is copied many times over the network, this network is organized so as to continually update and adapt all copies so that all records are congruous. No single machine or entity is in charge for the blockchain. The property of not having a central storage location makes manipulation or corruption almost impossible. Beginning with the basic record, all modifications added to the ledger can never be changed or altered. As a result, each time a copyrighted file is illegally utilized, a digital ledger storing the information from the owner and transactions history is public and simply verifiable.

For instance, Binded [16] defines itself as "the world's first copyright platform" for blockchain, giving rise to "a unique fingerprint" (cryptographic hash) for each copyright record. By tracking copyright records, Binded eases the protection of copyright by allowing the access to the circulation path of copyright knowledge by means of the blockchain. With a fingerprint at their fingertips, photo owners can monitor the online sites that use them. A demonstration of the attractiveness of this technology by large companies in the industry was given by Spotify. The multinational company, leader in music streaming, bought a start-up called "Mediachain Labs" [17] which developed a decentralised database of multimedia content and an automatic attribution engine. When the system recognizes the use of content registered on the platform, it automatically remunerates the right owner with a dedicated cryptocurrency. Another very interesting platform is Monegraph [18] which, in addition to allowing anyone to register their creative work on Bitcoin's blockchain, offers a number of additional services that allow users to avoid having to rely on intermediaries to manage the sale of licences. Through this system, in fact, the rights holder can freely set prices and conditions and, at the same time, receive payments.

However, there are also limitations to copyright protection such as initial authentication when uploading a file, such as proving ownership when a user uploads something. Other limitations, such as monitoring issues, the use of anonymous and untraceable buyer licenses are still to be resolved.

1.2.6 Voting

Voting is a fundamental element of any collaboration between people, with methods ranging from a show of hands to the ballot box, to new online voting systems. Traditional voting systems present various problems both in terms of security and cost. Just think of the system of balloting that can be manipulated to make the vote take the desired turn and the cumbersome system of counting the votes that requires a lot of labor and can not always be considered transparent as voters must trust electoral commissions, regardless of the high cost of elections. Internet voting (e-voting or online voting) allows voters to transmit their votes remotely from any computer connected to the Internet, potentially from anywhere in the world. Managing voting through online platforms is not an easy task, as on the one hand it is necessary to guarantee the privacy of the voter, both in terms of identity and preference expressed by the vote; on the other hand it is necessary to be able to have a certain degree of verifiability, i.e. to make the vote expressed by the voters represent their exact preference and above all is not manipulated.

Online voting systems save a lot of money for elections because voting and counting activities are digitalized, but security issues can arise. The security of an online voting system is necessary to ensure the reliability of the result and the confidence of the voter in the voting system itself. Trust requires sufficient transparency to allow the voter to accept the results of the vote without reservation. An e-voting method must therefore be developed with great consideration for certain key aspects of computer security and data protection. First of all, it is necessary to guarantee the authenticity of the preferences expressed by voters in order to ensure an exact final count. To achieve this goal, it is necessary to ensure that each vote is recorded without any possibility of change and to eliminate the possibility that voters may exploit imperfections in the system to vote more than once, while remaining anonymous. Therefore, e-voting systems should be robust and reliable enough to ensure the exact correspondence between the votes expressed and the votes recorded, despite the occurrence of possible network communication losses, server failures or cyber-attacks.

The use of blockchain technology could solve most of the problems related to both traditional voting systems and e-voting systems. First it is possible to solve the problem of transparency of the counting process, as with the blockchain you can track votes, thus ensuring transparency and protecting privacy as transactions remain anonymous. Regarding the e-voting systems, using the blockchain would be impossible for a user to vote twice, as on a chain. Also thanks to the immutability of the blockchain, once the vote is inserted in a block and becomes part of the chain is impossible to remove or modify it. The whole voting process would be transparent, secure, fast and much cheaper than a traditional voting system.

Today there are several projects that offer online voting systems based on blockchain. Among the most famous are Polys and Crypto-Voting. Polys [19] is an online voting project developed by Kaspersky innovation hub based on blockchain that can be used by companies, universities and political parties. The system is based on the use of devices connected to each other through a blockchain network that allows you to transmit and process together, securely, all the votes: both those at a voting station using the Polys Voting Machine and those released on their own devices. Voters are authenticated with documents that prove their identity and receive a unique Qr code of which only the voter is aware. This code is necessary to access the voting system and to check, through a special web application, that the vote has been registered in the blockchain. For reasons of privacy and secrecy of votes, the name and choice, are not stored in the blockchain. The access code for voting can be related to a poll in a specific field. This means that at the time of voting, the user is shown only the representatives for that vote specified even if the user is in a voting station located in another country. In addition, the automatic counting of votes significantly reduces personnel and resource costs for the organizers, allowing the results to be available as soon as the election is over, making the whole process much faster and more efficient. Crypto-Voting [20] is an original and integrated digital voting system relying on Blockchain technology, that was set up by Sardegna Ricerche under the POR FESR Sardegna 2014–2020 and implemented by a joint project from Net Service S.p.A. and the Department of Mathematics and Informatics of the University of Cagliari. The project aims to implement an online voting platform through the use of two concatenated permissioned blockchain: one to register the voters entitled and record the voting operation of the voters and another to count the votes assigned to the various candidates. The main purpose of the platform is to allow voters who are away from home or living abroad to vote remotely through mobile devices or personal computers. The voting operation consists of marking a symbol in a virtual ballot paper shown on the screen of your device. The voting process is divided into three phases, the first two, preparation of electoral lists and voting management, are performed on the first blockchain, while the third and final phase, the counting of votes, is performed on the second blockchain. This provides a transparent and public double ledger containing all voting results, which can be verified at all stages of the voting process.

1.2.7 Energy

The continuous growth of energy production from Renewable Energy Sources (RES) distributed throughout the territory has led to a change in the structure of the traditional electricity system and consequently of the energy transaction model. The electrical system was born as a centralised system, with power flows from production plants transferred through High Voltage transmission lines to Medium and Low Voltage grids to the end users. The advent of distributed generation and small domestic production plants has led to a change in both the electricity system and the energy transaction model, from centralised to distributed, with power flows that are inverted on the distribution networks and with energy input to the grid at all voltage levels. Both distributed generation plants and small plants are decentralised and widespread in the territory, these, thanks to technological innovation, have undergone a fast evolution. Especially for domestic plants, production was initially intended for self-consumption while the surplus was fed into the grid. With the advent of storage systems, the unused part remained destined to the grid, but it is a share that has started to reach important levels, both for the growing number of producers and for the greater efficiency of the plants. In addition, the intermittent production from renewable sources, if not optimally managed, can compromise the stability of the electrical system, as it may result in unbalances and possible congestions. One possibility to solve the issue could be the storage of this energy; another is the balanced Peer-to-Peer (P2P) exchange of energy between end-users and prosumers. The traditional energy transaction model, where each transaction is mediated by a third party, does not comprise the direct exchange of energy between end-users and prosumers. The lack of digital platforms supporting this business model is one of the reasons of its limited implementation. Thanks to the intrinsic features of the blockchain technology, namely decentralization, immutability, trust, transparency, traceability and total distribution, it appears to be the solution for such implementation and for the solution of many other problems related to the implementation of innovative business models. The massive advent of Renewable Energy Sources has transformed the electrical system from a centralized infrastructure to a distributed one. The intrinsic features of the blockchain can support the electrical system transformation to a fully decentralized system and enable new business models for the energy market. As a result, currently many projects in the energy area are adopting the blockchain technology as main solution.

Looking at the literature, the applications of the blockchain in the energy field can be categorized into two main classes:

- *Electricity trading*: covers all applications for which two parties exchange electrical energy with a unit of value, such as ir happens in P2P energy trading. These can be managed centrally by a network operator (DSO or TSO) who will verify the correspondence of the exchanges in order to obtain the maximum benefit in technical terms for the network or they can be managed in a distributed way, as in some cases the constraints of the network are not so rigid.

- *Renewable energy certification* and *demand traceability*: includes applications related to the certification of energy from renewable sources and the traceability of demand/response aggregation. In this case, the blockchain platform can register the green energy injection in the network at a certain node owned by a producer or certifies the participation of a certain prosumer in an aggregation or demand response program, enhancing the transparency of the certification of green energy or the remuneration mechanisms of demand response services.

The applications that are possible are numerous, but the most popular and promising are: P2P energy trading, Vehicle-to-Grid (V2G) and Demand-Response (DR).

Before the advent of blockchain was difficult, if not impossible, perform energy transactions between end users in a safe and transparent way, as the market and the electricity system were structured so that energy transactions were centralised and managed by trusted entities. The blockchain, thanks to its features and advantages, has become very famous for enabling the trade of P2P electricity between small producers and consumers, especially within the microgrid.

Blockchain technology also offers many advantages when applied to automotive systems. With this in mind, the so-called V2G applications have been developed in recent years. V2G represents a system for management, storage and distribution of energy by batteries of electric cars, allowing also to guarantee their diffusion at more convenient prices. Thanks to this technology and the integration between electric vehicles and the grid, owners have the possibility to use their car as a mobile power station integrated into the grid, so that the energy stored in the batteries can be transferred to the grid in case of demand or to other vehicles. In parallel to facilitating transactions between vehicles or vehicles and grid, the implementation of blockchain-based solutions could solve some problems related to the after-sales of the vehicles, such as the traceability of original spare parts, identification of counterfeit parts or the recovery of the vehicle at the end of its life in a circular-economy perspective. To date, it is virtually impossible to find this information because it is either inaccurate/incomplete at the time of transmission or has been modified by hackers. In this sense, the blockchain offers a more secure and transparent way to communicate thanks to the continuous sharing of information between nodes of the same and the increased difficulty of tampering with individual blocks.

In addition to P2P energy exchange and V2G applications the blockchain offers the possibility of allowing the exchange of energy services rather than the direct exchange of energy between two users or user and network. To cope with sudden power fluctuations due to RES unpredictable behavior and massive penetration into the electrical system, DR programs are a very effective solution. In general, DR is a structured program that aims to change the load diagram of users in response to problems on the grid, such as congestion or temporary unavailability of electricity caused by failures or intermittent production from non-programmable RES.

Using the blockchain, DR programs can be run directly between the network operator and users, unlike the current scenario in which the aggregator mediates between the user and the network operator. In this case: the TSO/DSO requires a reduction/increase of the load and users respond accordingly by reducing or increasing the load. The transaction is not a direct energy provision from a producer to a consumer and subsequent payment (the two parties of the transaction), but rather a request to provide a service, after which users are remunerated if they respond adequately. One of the main problems in the issueing of traditional DR service is the lack of transparency, but through the blockchain it is possible to manage this problem. In addition, the blockchain allows small prosumers to participate in capacity and balancing markets by aggregating them into virtual load units that can be managed as needed.

These three applications, as well as the Energy Communities introduced by the European Directive on the use of renewable energy sources, are new ways of organising energy resources, necessary to achieve the objectives of energy savings and RES penetration set for 2030.

More details on Energy Communities are given in Chap. 2, while the P2P, V2G and DR applications and their experiences are outlined in Chap. 3.

1.3 Blockchain Technology Fundamental Principles

1.3.1 Distributed and Centralised Systems

The data management computer systems, based on the place where the database (DB) is located, are divided into centralised and distributed.

A computer system is called centralised when the DB containing the data and applications is located on a single processing node. The DB can be local if it is located on the same computer of the user, or remote if it is located on another computer or server where one or more users can access remotely (see Fig. 1.3a). Depending on the mode of operation, centralised systems are divided into single-user systems, in which services are aimed at one person at a time, or multi-user systems, when both data and management application programs are placed on a server accessible from client computers through a network of computers (client-server architecture). While, a system is called distributed when the applications, between them cooperating, or the DB are located in two or more computers connected in a local or geographical network. In general terms, a distributed system consists of a set of logically independent applications that work together to pursue common goals through a hardware and software communication infrastructure. The DB can be located in several computers located in the same place, or distributed in a network of computers connected to each other in the form of a distributed system (see Fig. 1.3b). The data on a distributed network are data with information constituting a single logical structure but physically stored on several autonomous computers and connected through a network. The location of the data is transparent to the users' applications and the information is available for use by different applications and users in the network of computers, regardless of the geographical location where the data is stored.

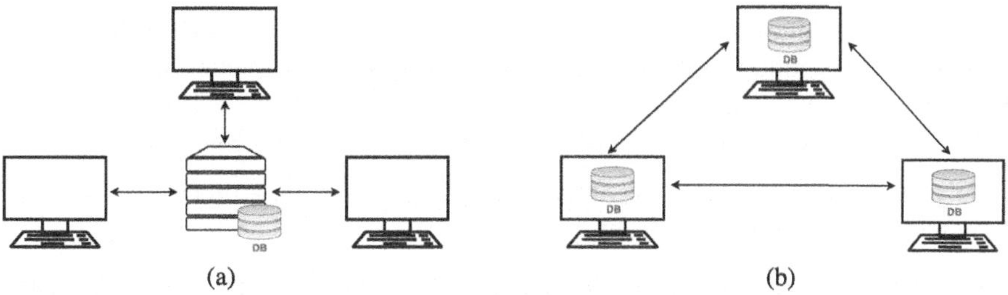

Fig. 1.3 Centralised (**a**) and distributed (**b**) networks

The two main advantages of centralised systems are:

- *High security*, since the database is managed by a single central authority that controls "who can do what" in the system.
- *Data protection*, since the data and programs all reside on the same DB, one update applies to all, one backup applies to all.

On the other hand, although efficient from a data protection point of view, centralised systems are characterized by several disadvantages:

- *No transparency*, since the DB is governed by a central authority that can theoretically modify the data.
- Applications are executed on a single process that represents the only operating part of the system, so there is *one single point of failure*.
- *Slow execution* and *poor interactivity* depending on the number of users connected to the central computer at a given time.

In general, the centralised systems are more exposed in the event of a cyber-attack, but they are good when the party that controls the system is honest and behaves reliably.

In distributed systems, each processing node has the same operating capabilities and the same permissions for access to whole information. This comes up with a robust foundation for trust, as it democratizes the entire system. Additionally, as data is stored on all the nodes of the platform, in the event of a cyber-attack, there is no single point of failure, so the system can keep on operating. Distributed systems, and thus also the blockchain, are adequate when there is a plurality of administrative authorities and there is a lack of trust between them [21]. So the main advantages are:

- *Transparency*, since there is no authority managing the system and all data is accessible and owned by all users.
- *Reliability* and *fault tolerance*, because thanks to the redundancy of the process nodes in case of failure of one of them the system will continue to operate.
- *Scalability*, since as the requests increase, the system can be expanded with simplicity because high computing power is not required for individual nodes, and the more nodes there are in the network the more resources are made available.

- *Economicity*, since the computing power is distributed among the different processing nodes, rather than concentrated on a single node as for centralised systems.
- *Integration*, since it allows to connect different machines, even different ones, as long as they are able to interface to the same communication subsystem.

Distributed systems therefore allow to solve problems that afflict centralised systems, but they also have disadvantages, such as:

- *Complexity*, since it is needed to manage several nodes at the same time rather than a single node.
- *Security*, since it is difficult to control access, the distribution of data is intrinsically insecure and each user may theoretically modify some data.
- *Communication*, because in order for a distributed system to be efficient, high standards of communication are required.

The blockchain technology is a DLT, so a technology that is based on the distribution of the database over multiple nodes but that allows to overcome the security problems that afflict distributed systems through the use of encryption. Security is achieved through the use of consensus algorithms that are more or less complex depending on whether the blockchain is permissionless or permissioned. In permissioned blockchain, a new user needs the prior approval of an entity before joining, while permissionless blockchain allows anyone to participate in the network. In permissioned blockchain, the identity of participants is known, while in permissionless blockchain participants are anonymous. Data security and trust between participants is achieved by using the consensus algorithms. A consensus algorithm can be defined as the mechanism by which a blockchain achieves consensus on the validity of transaction blocks and thus the data recorded on the blockchain. In a permissioned blockchain they are lighter as users need a permission to participate in it and therefore they are known and theoretically trusted. While the consensus algorithms are more computationally intensive in a permissionless network because users are anonymous and the consensus algorithm has the task of ensuring trust between users. Besides, blockchain largely uses cryptography to keep the register secure and immutable, while smart contracts are employed to execute or validate transactions. Below the consensus process with the most widely used algorithms and smart contracts are better explained by referring to blockchain both permissionless and permissioned.

1.3.2 Consensus Algorithms

"Consensus" consists of finding an agreement on the validity or truthfulness of something. In this case, the consensus is reached on transactions and, thus, on the blocks of transactions that can be inserted in the blockchain. This way to operate allows to do without a third party that is instead a basic element in traditional centralised systems. It can thus be said that consensus is the most important part of the blockchain technology implementation. In blockchain applications, consensus on the validity or truthfulness of transactions, and therefore blocks of transactions, is achieved by running consensus algorithms. The latter are more or less computationally intensive and are based on different mechanisms such as solving a puzzle or exploiting the outcome of a voting session. Consensus algorithms must also ensure the so-called Byzantine Fault Tolerance (BTF), namely they have to guarantee that the distributed system keeps on working even if some nodes operate incorrectly or act not honestly. The complexity of the algorithm is different if the blockchain is permissioned or permissionless. Usually, these algorithms require heavy computations in the permissionless blockchain, as users' identity is unknown and therefore the consensus algorithms are necessary to ensure that the users are not malicious and to ensure the consistency of the data stored on the blockchain. Permissioned blockchain use lighter consensus algorithms only necessary to ensure the integrity of the data exchanged, as users' identities are known and usually there is no need to verify that a user is malicious. The most known consensus algorithm ensuring the BTF is the Proof of Work (PoW), used in Bitcoin [1]: in this case, the user who adds a new block on the chain is able to solve a computationally-intensive problem before the others. In the case of Bitcoin, some users compete for the insertion of new blocks and they are called "miners"; the system pays the effort made to solve the algorithm to these "miners" a fixed number of new bitcoins. As Bitcoin is supported by a permissionless blockchain, in this system, any user owing a network node can run the consensus algorithm and compete with other users on other nodes for the insertion of new blocks. In this way, they will earn both the bitcoins issued by the system as a payback for the computational effort and the bitcoins that users pay to miners to have their transaction inserted into a new block. It is easy to understand that the higher the computational power, the higher the probability of succeeding in the block's validation, but as the cryptographic puzzle is resolved and a new valid block is identified, the majority of the nodes of the entire network must approve it. If this does not happen, the block will be refused and the miner will not earn bitcoins. Moreover, since the network can grow in size along time, to keep high the level of security, the energy consumption required to implement the consensus grows too. In theory, a user who is malicious could include a false transaction inside a block and validate it. However, following the consensus process as outlined above, the block would not be approved by the majority of nodes and would thus be discarded. Due to the excessive energy consumption for the execution of the algorithm, to date and on average, about 1794,60 kWh per transaction [22] are spent. This amount of energy is equivalent to the power consumption of an average U.S. household over 61.51 days. As the network grows on the other hand it is more and more difficult for a malicious user to own or control 51% of the network nodes. As a result of the application of this mechanism, not only malicious users get eliminated from the system, but also 'miners' are compelled to validate the blocks correctly. Indeed if these latter do not act correctly, they would not be remunerated and face high costs in terms of energy consumption.

As an alternative to PoW, consensus can be reached with much less computational effort and thus energy consumption by means of the Proof of Stake (PoS) consensus. Differently from PoW, for which the algorithm rewards 'miners' who are capable to solve the cryptographic puzzle so as to validate transactions and create new blocks, using PoS, the creator of a new block is identified based on its wealth, also termed "stake". The higher the stake, namely the amount of digital currency belonging to a user, the higher the probability that the system is not being hacked. Additionally, if an individual is exposed to a given cryptocurrency running on a given platform, it is likely that he or she is behaving honestly. In blockchains using PoS, there is no reward for creating new blocks and all digital currencies are generated at the start and their number never changes. Indeed, in these platforms, the miners are also named "forgers". In these systems the transaction fees are used to remunerate the generation of a new valid block. In blockchains using PoS, participants with a large amount of cryptocurrency are selected on a pseudo-random basis to coin or 'forge' new blocks and add them to the blockchain. The selection process begins after the analysis of several factors and on a pseudo-random basis, so as to ensure that not only individuals with a higher stake are selected, but also others with a lower stake. By now, Ethereum's blockchain adopts PoW for blocks' validation, nevertheless, for some years its creators have planned to switch to PoS with a protocol called "Casper" [23] since the PoW process is very energy intensive. The energy costs are for now paid in fiat coins, leading to continuous downward pressure on the value of the related digital currency. For this motivation, Ethereum developers have been trying to leverage PoS for a "greener" and thus less expensive distributed consensus. It must be mentioned here that one of the weak points of PoS is security. In a system based on PoW, the cost of executing an attack is really high, as an hacker would risk spending more than what he would earn, thus malicious users are left out. A system based on PoS might be cheaper to attack. Due to this reason, the Casper protocol provides in some cases for the loss of deposit that the (malicious) user has set aside to take part to the creation of new blocks and have a privilege to be considered as part of the consensus mechanism. Alternatively, it would be possible to grow the value of the cryptocurrency, thus increasing the cost of attacks. The type of consensus algorithm, and related complexity, strongly influences the blockchain technology as the nature and strength of the distributed consensus algorithm affects the security of the blockchain. The management model of consensus influences the difference between permissionless and permissioned blockchains. In a permissionless blockchain-based platform, trust among anonymous users is secured by the consensus algorithm. Differently, in a permissioned blockchain, identities are known to all platform members and permissions to operate are released by an administrator of the network. In this latter case, it is possible the use of a lighter consensus algorithms. In a permissioned Blockchain platform, not only permits to operate are released by a central authority, but even permits to access the same platform and become a member. In such case, as roles of different users are determined by the administrator and identities of users are known, also validators are known. As a result, there is no risk of a 51% attack (from the majority of users) resulting from miners collusion. For this reason, transactions are not as energy intensive and are cheaper, as they are verified only by a small number of nodes, named validators, which are to be considered 'trusted'. When the consensus is carried out by a set of known and reputable users/nodes, the consensus mechanism is called Proof of Authority (PoA) [24]. Validators are member nodes that are authorized to take part to transactions' validation and new blocks insertion to the blockchain. The fact that the identity of the 'miners' or validators is known means that their work does not have to be controlled and verified by other nodes, thus reducing the burden, time and cost of execution of the consensus process. In this case, trust does not stem from the consensus algorithm used, as simpler and faster algorithms can be employed in permissioned networks. The most commonly used algorithms are the practical-Bizantine-Fault-Tolerance (pBFT) ones. These have limited complexity and high applicability in distributed systems. It is proved that in pBFT systems, the maximum number of malicious nodes that do not put at risk the correct execution of the validation process is 1/3 of all system nodes in a given vulnerability window. As the number of nodes grows, the security of a pBFT network increases. When a user sends a request for transaction, the latter is received by a validator node and one of the validators is randomly chosen as "leader". As a new block is created, the "leader" sorts the transactions that are expected to be included in the block and sends the list of transactions that are sorted to all the validator nodes in the network. After the reception of the list, each validator starts executing the transactions one by one. After this phase, each validator calculates the hash code for the block and sends it out to the "leader". The latter then counts the responses. If he acknowledges that 2/3 of all validators have the same hash code, he will commit the new block on the blockchain.

In permissioned blockchain technology, validators do not compete to find a solution, such as it happens in permissionless environments and PoW-based consensus, and this translates in more environmentally friendly solutions. The PoW consensus indeed is based on a competition for the solution of a difficult cryptographic puzzle. Generally, each consensus algorithm has pros and cons; each of the features can be valuable for certain applications and not for others. What can be said in the energy field, in which energy consumption is considered a main issue in all applications, is that energy intensive consensus algorithms should be avoided.

1.3.3 Smart Contracts

When two or more parties want to establish, regulate or terminate a relationship of any kind, a contract can be deployed. In this era of digital transition, the blockchain technology offers the possibility to virtualize the conclusion of contracts, as it offers many advantages in terms of transparency, information symmetry and disintermediation. One of the features of certain blockchain platforms is the possibility to implement the so-called "Smart Contract" (SC). A SC can be defined as the "translation" into a code (virtualization) of a contract between different parties. The codification of the contract allows to verify automatically specified and predefined conditions and carry out actions when conditions between the parties are fulfilled and verified. A traditional contract is composed of different parts and namely: the agreement between the parties, the motivation for which the contract is deployed, the object and the form, when it is required a specified form (written/oral). Smart contracts are based on the logic of "if-this-then-that", that is to say that when an assumption (this) foreseen within the contract occurs, a result (that) follows. These contracts therefore have the ability to enforce their clauses and to have prompt and immediate execution without the support of an external party. In blockchain, the agreement between the parties and motivation in this case appears after the 'consensus' of the platform users on the code implementing the SC, while the object comes after the logic implemented by the 'if-then' rules. As an example, if a clause stipulates that a party in the event of a performance will be paid automatically from the account of the defaulting party, the sum agreed as a penalty will be automatically paid from the account of the defaulting party to that of the complying party. As a result, it guarantees the same trust as a traditional contract, but with the advantage of automatically verifying the terms of the contract, performing actions such as money transfer and assessing the situation, if one of the parties does not comply with the terms. However, immutability is seen as a weakness in most economic sectors as, especially as the time goes by, the contract keeps being executed under conditions that may no longer be valid. SCs deployed by a blockchain are shared on network nodes and are usually executed to implement a given transactions logic and verify its validity. They also execute transactions automatically, if certain conditions are met. Another important feature of SCs is that no one in the network has whole power over the contract. Additionally, as they are implemented on the blockchain, they inherit some interesting features:

- *Immutability*: after the creation, a SC can never be altered. No one can modify the contract's code, not even the creator of the contract itself.
- *Transparency*: each blockchain user has visibility over the SC's code and can verify its validity.
- *Trust*: the contract is distributed on the network and validated by the blockchain. A single node cannot force the contract against specified features, as the other nodes in the blockchain would not approve it.

The SCs were introduced the first time with Ethereum [25] technology, a worldwide and open-source permissionless blockchain for decentralised applications that implement SCs for cryptocurrency transactions execution. In this case, Ethereum SCs are written in a programming language called "Solidity". The latter is a Turing complete language, which is unsuitable for complex codes implementation, such as the code for the solution of the optimal power flow problem in power systems or other optimization problems. Ethereum SCs can be compiled in "bytecode" and can be read and executed by an "Ethereum Virtual Machine" (EVM). An EVM runs on each node of the Ethereum network, and the EVM is the environment in which SCs are executed. When a transaction must be written within the Ethereum blockchain, a SC is executed and, after that, a fee or remuneration in "gas" is paid for this execution. The quantity of "gas" for such remuneration is directly proportional to the length and computational complexity of the SC to be executed. It is further possible to define the so-called "gas price", i.e. the amount of Ether (the Ethereum cryptocurrency) per unit of "gas" anyone is willing to pay. The alteration of such value, influences the priority of the transactions to be inserted in the blockchain, in fact, as the value of the "gas price" gets higher, the transaction will be recorded in a shorter time as compared to a transaction implemented with a lower value of "gas price", since the node who

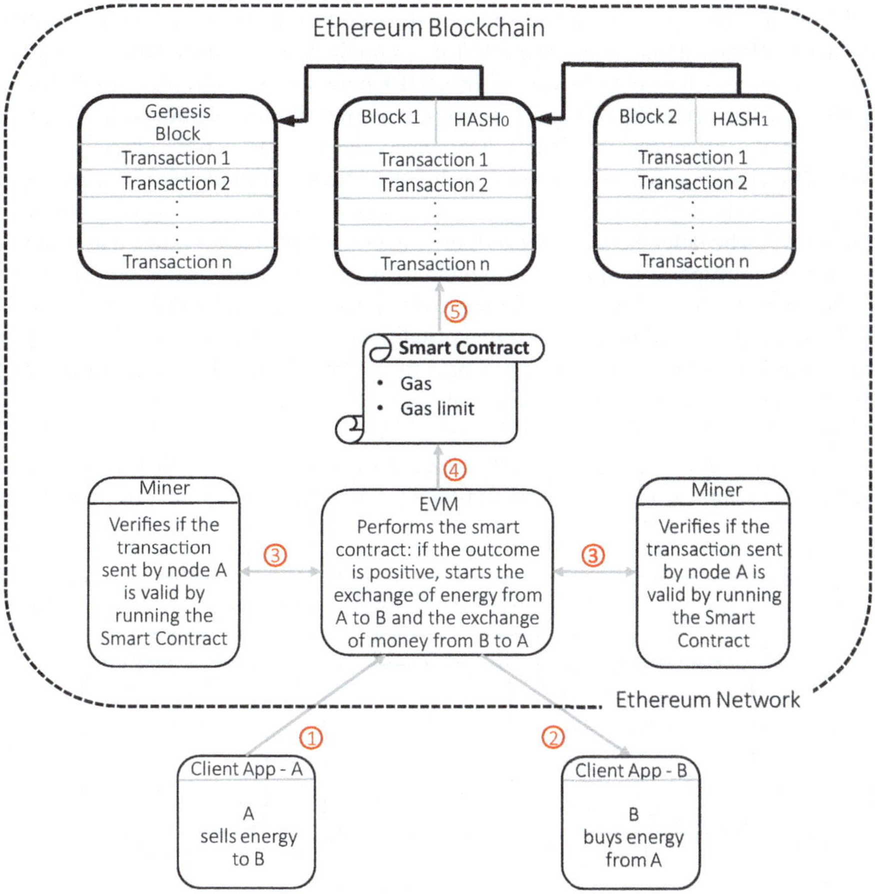

Fig. 1.4 Ethereum smart contract implementation

executes the smart contract for inserting the new transaction will earn a higher fee. As a result, there is an increase in the speed of registration of the transactions, when the price is higher. Figure 1.4 shows the way in which Ethereum SCs are implemented referring to the application of Peer-to-Peer energy exchange between two users.

As already said, Ethereum is a permissionless blockchain. The use of permissionless blockchains gives greater security, when data can be released publicly, but, as there is a need for privacy and for limiting the computation time in case of heavier consensus algorithms, this technology may not be adequate for all applications.

A very popular permissioned blockchain that implements SC is Hyperledger Fabric, HLF, [26] from IBM. HLF is a platform conceived for business contexts applications, with a highly modular and configurable architecture that supports innovation and versatility for a wide range of industrial use cases. Fabric, unlike Ethereum, enables SCs written in general purpose programming languages such as Java, Go and Node.js. The SCs are named "chaincodes" and they operate as a reliable distributed application that attains security/trust from the blockchain and the underlying consensus among the network's peers. In an HLF network, the SC represents the main element of the network, because it is the component that implements the logic of each kind of transaction and verifies the integrity of each transaction sent by network users. The chaincodes allow the users to directly interact with the blockchain ledger and, as a result, any operation, even a simple query, is performed in HLF through chaincodes. The latter are installed and running on each peer. Every time a user sends a transaction proposal to the platform and this transaction proposal is processed by the SCs running on all peers. If the transaction sent by the user is consistent with the logic implemented by the SC and verified by all peers, the new transaction after the consensus process is recorded in a new blockchain block. The entire process to register an energy transaction as well as all the components of the network are shown in Fig. 1.5.

Summarizing, it can be said that the SCs are software programs that let the transfer of ownership of any digitally identifiable good from one person to another or

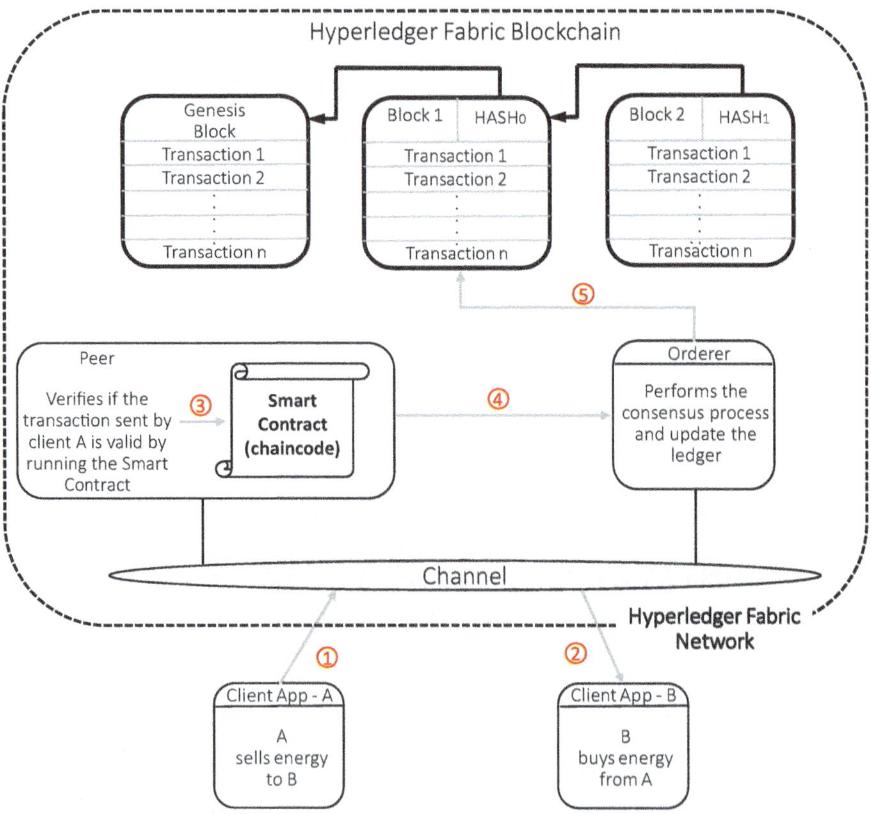

Fig. 1.5 Hyperledger fabric smart contract implementation

the implementation of a given transaction logic for a particular application. As an example, in P2P electricity transactions, the SC can be used to execute energy transactions directly when the conditions for trading set by the prosumers are met, or to assign to the prosumers digital tokens corresponding to the energy injected into the grid and destroy them as the user consumes electricity [27]. Tokens are delivered by the blockchain platform. They can either represent, as in this case, the digital representation of an asset or they can be used as a payment (payment token), or they can even demonstrate the ownership of a right (service or utility token). The latter right can be either connected to the ownership of a financial asset (service token) or to the likelihood to obtain a service/benefit (utility token) by means of the platform. In Demand Response, DR, programs, tokens can be used to implement the logic of Demand Response to certify the energy services provided by users in return for remuneration [28]. In Vehicle-To-Grid, V2G, applications, SCs can certify the ancillary service offered by an electric vehicle during charge or discharge [29].

1.3.4 Blockchain and General Data Protection Regulation

In the previous paragraphs, it has been explained how a blockchain works, its characteristics, types and different consensus processes, but the protection of personal data of users participating in it has still not been mentioned. The General Data Protection Regulation (GDPR) became effective on 24 May 2016 and applied in EU countries from 25 May 2018, with the aim of providing guidelines for the protection of personal data in computer systems, including the blockchain [30]. The GDPR introduces a system of governance of personal data based on a responsible figure called "Data Controller", who is also the data controller. The latter must prevent and correct any errors and must also be able to demonstrate compliance with GDPR to the authority, for example through internal documentation or policies. The innovation but also the problems of the blockchain arise from the fact that it proposes to transmit and store information on an immutable decentralised register, validated not by a traditional central authority (such as a bank or government) but by the "public" itself. The core of this technology is the decentralization, transparency and immutability of data, concepts that seem contrary to the directives of the GDPR.

The GDPR provides the six essential principles for data protection:

1. treatment must be lawful, fair and transparent;
2. the processing of data must be limited to the specific purpose for which they were originally collected (purpose limitation);
3. it is possible to collect only the data absolutely necessary for the specific purpose (data minimization);
4. data must be accurate and up-to-date (accuracy);
5. data must not be kept longer than what is needed (retention limitation);
6. data must be processed securely (integrity and confidentiality).

Most issues in the compatibility between GDPR and blockchain relate to permissionless blockchain, such as Bitcoin, which are open, without a centralized ownership and designed not to be controlled. In this case, indeed, all participants can act as validators. Conversely, permissioned blockchains, although decentralised, allow access to a limited number of users, who obey to strict rules and hold given features. The GDPR was created to be applied to a centralised context, so one of the first problems to find compliance is the difficulty of identifying a Data Controller in the blockchain. A direct consequence is that in case of dispute, there is no competent jurisdiction, since there is no one who is responsible of the data. Then, according to the fifth principle of GDPR, the data must be kept for the time that is strictly necessary to achieve the purposes for which they are processed. In particular, the interested party must have the right to request that their personal data, that are no longer necessary for the purposes for which they were collected, can be deleted and no longer processed. However, the blockchain does not let data cancellation, once written in the chain. Finally, according to the GDPR, the transfer of personal data to a third country or an international organization is permissible if the European Commission has decided that the third country, a territory or one or more specific sectors within the third country, or the international organization in question guarantee an adequate level of protection. In permissionless blockchain, it is complicated to verify these guarantees, as users from all over the world can participate. On the other hand, in permissioned blockchain, it is possible to surmount this problem by means of the use of specific clauses. Differently from how it may seem at a first glance, thanks to the characteristics of immutability and replication, the blockchain can play a key role in GDPR. If designed correctly it is able to ensure transparency and control of personal data. In fact encryption and a decentralised structure make the blockchain network very resistant to tampering and its transparent nature offers clear access to data in line with the objective of returning control and visibility of the data. The blockchain offers a radically different approach to computer security, which can go so far as to certify the identity of a user, ensure the security of transactions and communications, while protecting the infrastructure that supports operations between organizations: a paradigm shift that allows to take full advantage of shared online services. Regarding the ownership of personal data, there is the possibility to choose different possibilities. The use of permissioned blockchain implies the presence of a network administrator, which is also responsible for users identities management. For example, in the energy sector the role of administrator of the network is different depending on the application: in P2P, it could be represented by the DSO, as it is in charge of the correct operation of the Low and Medium Voltage grids; in V2G applications, the role could be taken by the aggregator managing the EV charging stations and "selling" capacity to the system operator; in DR programs, the same role could be taken by the Transmission System Operator, TSO, as it is responsible for regulation in power systems; finally, in energy communities, the administrator could be the same building administrator or the head of the community. More complex solutions provide the possibility to store personal data outside the blockchain, registering only a link to those data within the ledger. So, the blockchain would only be used to store the evidence that some data exist rather than to store the data them-

Table 1.1 Compliance between GDPR requirements and blockchain

GDPR requirements	Permissioned blockchain	Permissionless blockchain
Data controller identification	The Data Controller can be the network administrator or a user chosen for the purpose	Difficult to identify, each user can be a validator
Competent jurisdiction identification	The administrator can represent a competent jurisdiction in case of dispute	Difficult to identify
Data retention limitation	It does not allow to delete data once written in the chain, but there is the possibility to store on the blockchain the link of data stored externally	It does not allow to delete data once written in the chain, but there is the possibility to store on the blockchain the link of data stored externally
Identification of a commission for the transfer of personal data to a third country	The problem can be overcome through the use of contractual clauses or certification mechanisms	Difficult to track the location of the participants
Data transparency and security	Encryption, decentralised structure and immutability ensure security and transparency of personal data	Encryption, decentralised structure and immutability ensure security and transparency of personal data

selves, thus allowing the removal of personal data without breaking the chain. Some research projects are now analysing the use of particular hash functions that add an electronic lock between the different blocks, giving the administrator a digital key to unlock and modify them [31].

Table 1.1 summarize the main requirements of GDPR and how the use of blockchain technology (permissioned) allows to fulfill them.

Having assessed the compatibility between blockchain and GDPR, other regulatory aspects remain open. In fact, in Italy, several projects define new business models and propose improvements to existing ones, but in order to move from the definition phase to the market one, a legislative step is required to extend the recent legislation on energy communities and V2G recognizing the role of new technologies in the technical management and economic of electricity distribution networks. These topics will be discussed in more detail in the next chapter.

References

1. S. Nakamoto, *Bitcoin: A Peer-to-Peer Electronic Cash System* (2008). Accessed 12 May 2020. [Online]. Available: https://bitcoin.org/bitcoin.pdf
2. ENTSO-E, *Towards Smarter Grids: Developing TSO and DSO Roles and Interactions for the Benefit of Consumers*. Accessed 12 May 2020. [Online]. Available: https://bit.ly/entso-e

3. eex group, *EPEX SPOT markets: Focus on Flexibility* (2019). Accessed 13 May 2020. [Online]. Available: http://powerswarm.co.uk/wp-content/uploads/2019/01/2019.01.23-Wright-EPEXSpot-Flex-Markets.pdf
4. S.K. Lo, X. Xu, Y. Chiam, Q. Lu, *Evaluating Suitability of Applying Blockchain* (2017), pp. 158–161. https://doi.org/10.1109/ICECCS.2017.26
5. X. Xu, I. Weber, M. Staples, Design process for applications on blockchain, in *Architecture for Blockchain Applications* (Springer International Publishing, Cham, 2019), pp. 93–111. https://doi.org/10.1007/978-3-030-03035-3_6
6. G. Foroglou, A. L. Tsilidou, Further applications of the blockchain, in *Conference: 12th Student Conference on Managerial Science and Technology, At Athens* (2015)
7. W. Chen, Z. Xu, S. Shi, Y. Zhao, J. Zhao, A survey of blockchain applications in different domains, *ACM International Conference Proceeding Series* (2018) pp. 17–21
8. J.R. Varma, Blockchain in finance. Vikalpa **44**(1), 1–11 (2019). ISSN: 23953799
9. B. Wu, T. Duan, The application of blockchain technology in financial markets. J. Phys. Conf. Ser. **1176**, 042094 (2019). https://doi.org/10.1088/1742-6596/1176/4/042094
10. H. Dalianis, A. Henriksson, M. Kvist, S. Velupillai, R.Weegar, HEALTH BANK—A workbench for data science applications in healthcare, in *CEUR Workshop Proceedings*, vol 1381 (2015), pp. 1–18. ISSN 16130073
11. E. Scheuer, Health information traceability foundation, in *Whitepaper: HIT Foundation*, vol 2.1 (2019), pp. 1–22
12. A. Kamilaris, A. Fonts, F. Prenafeta Boldú, The rise of blockchain technology in agriculture and food supply chains. Trends Food Sci. Technol. **91** (2019). https://doi.org/10.1016/j.tifs.2019.07.034
13. G. Motta, B. Tekinerdogan, I. Athanasiadis, Blockchain applications in the agri-food domain: The first wave. Front. Blockchain **3** (2020). https://doi.org/10.3389/Blockchain.2020.00006
14. SNAPPlab and ENGinfo, *Seedsbit, Beyond Information Traceability* (2020). Accessed 15 Nov 2020. [Online]. Available https://seedsbit.com/en/
15. K. Sayegh, *Blockchain Application in Insurance and Reinsurance* (SKEMA business school, 2019)
16. BINDED: Copyright made simple. Accessed 19 May 2020. [Online]. Available https://binded.com/
17. Media Chain, *MEDIACHAIN: A De-centralised Blockchain Focused on the Film Industy*. Accessed 19 May 2020. [Online]. Available https://wxappres.feeyan.com/block/project/3710.pdf
18. Q. Dupont, Blockchain identities: notational technologies for control and management of abstracted entities. Metaphilosophy (2017)
19. R. Alyoshkin, *Polys Online Voting Systems*. Accessed 26 Aug 2020. [Online]. Available https://polys.blob.core.windows.net/site/Polys_whitepaper.pdf
20. F. Fusco, M. I. Lunesu, F. Pani, A. Pinna, *Crypto-Voting, a Blockchain Based E-voting System* (2018), pp. 223–227
21. M.J.M. Chowdhury, A. Colman, M.A. Kabir, J. Han, P. Sarda, Blockchain versus database: a critical analysis, in *2018 17th IEEE International Conference On Trust, Security And Privacy In Computing And Communications/12th IEEE International Conference On Big Data Science And Engineering (TrustCom/BigDataSE)* (2018), pp. 1348–1353
22. Digicomist, *Bitcoin Energy Consumption Index*. Accessed 8 July 2020. [Online]. Available https://digiconomist.net/bitcoin-energy-consumption
23. V. Buterin, D. Reijsbergen, S. Leonardos, G. Piliouras, *Incentives in Ethere-um's Hybrid Casper Protocol* (2019), pp. 236–244
24. V. Arasev, *POA Network Whitepaper* (2018). Accessed 8 July 2020. [Online]. Available https://github.com/poanetwork/wiki/wiki/POA-Network-Whitepaper
25. V. Buterin, *A Next Generation Smart Contract & Decentralized Application Platform* (2009). Accessed 9 July 2020. [Online]. Available https://blockchainlab.com/pdf/Ethereum_white_paper-a_next_generation_smart_contract_and_decentralized_application_platform-vitalik-buterin.pdf

26. E. Androulaki, C. Cachin, C. Ferris, C. Murthy, B. Nguyen, S. Muralidharan, M. Sethi, C. Stathakopoulou, "yperledger Fabric: a distributed operating system for permissioned blockchains, in *Presented at the Thirteenth EuroSys Conference 2018*, Porto, Portugal, 23–26 Apr 2018, no. 30
27. W. Tushar, T. Saha, C. Yuen, D. Smith, and H. V. Poor, "Peer-to-peer trading in electricity networks: An overview," IEEE Transactions on Smart Grid, vol. PP, p. 15, Jan. 2020. https://doi.org/10.1109/TSG.2020.2969657.
28. M.L. Di Silvestre, P. Gallo, E.R. Sanseverino, G. Sciumà, G. Zizzo, Aggregation and remuneration in demand response with a blockchain-based framework. IEEE Trans. Ind. Appl. **56**(4), 4248–4257 (2020)
29. H. Liu, Y. Zhang, S. Zheng, Y. Li, Electric vehicle power trading mechanism based on blockchain and smart contract in v2g network. IEEE Access **7**, 160546–160558 (2019)
30. Blockchain and GDPR. Accessed 1 Oct 2020. [Online]. Available https://www.smart4ius.eu/wp-content/uploads/GDPR-e-BLOCKCHAIN.pdf
31. E. Politou, F. Casino, E. Alepis, C. Patsakis, Blockchain mutability: challenges and proposed solutions. IEEE Trans. Emerg. Top. Comput. (1), 1–1, 5555. ISSN: 2168-6750. https://doi.org/10.1109/TETC.2019.2949510

Chapter 2
Regulations, Business Model and Actors in the Energy Market

G. Sciumè, E. Riva Sanseverino, and P. Gallo

Abstract The electricity market is a particular market with an inherent complexity. The mechanisms that regulate the movement of energy volumes are managed in a precise and systematic order to ensure transparency and competition on the one hand, and reliability and continuity of service on the other. The electricity market is made up of several well-defined phases, and in each phase one or more players can operate according to the purpose of the market phase itself. There are many players involved in the energy business, each with its own role and identity. In general they can be distinguished: *Prosumers*, *Aggregators*, *Energy Market Manager*, *Balance Responsible Party*, *Balance Service Provider*, *TSO*, *DSO* and *Energy Retailer*. As the electricity system evolves, the electricity market and the role of the different actors is also changing, enabling new business models that need to be properly regulated. This chapter aims to describe the various phases of the Italian electricity market and the actors involved in each phase, defining their functions and tasks, the new business models and the main regulations. Special attention is paid to the players involved in regulatory services and, consequently, to transactions connected to the balancing of the electricity grid.

2.1 The Italian Electricity Market

The electricity from the generation plants is transported to consumers through the power grid, an articulated system of cables and infrastructures. The power grid uses high voltages for transport over long distances to reach the consumer with lower voltage lines passing through substations. In the transmission phase, electricity from large thermal-electric power plants or from renewable sources (injection points) is injected in a High-Voltage grid, called the transmission grid. The interconnections of this grid are the electrical stations (SST) from which the distribution networks start, at lower voltage (medium and low voltage) that supply the consumption points connected to them. The distribution takes place through a first transformation of the

G. Sciumè (✉) · E. Riva Sanseverino · P. Gallo
University of Palermo, Viale delle Scienze, Building 9, 90128 Palermo, Italy
e-mail: giuseppe.sciume01@unipa.it

G. Sciumè et al. (eds.), *A Practical Guide to Trading and Tracing for the Energy Blockchain*, UNIPA Springer Series, https://doi.org/10.1007/978-3-030-96607-2_2

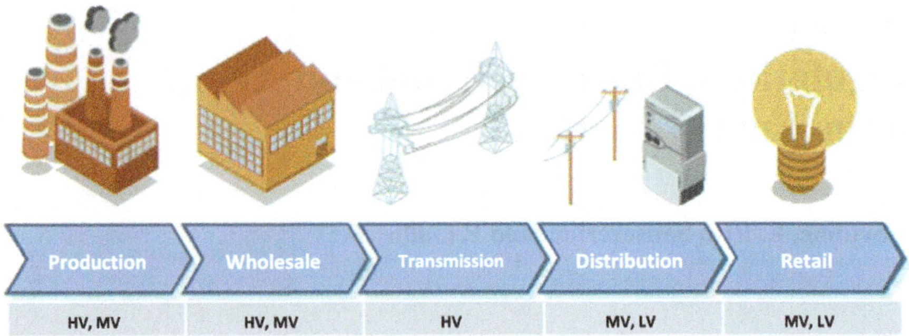

Fig. 2.1 Main phases of the electricity supply chain

medium voltage current (15–20 kV) starting from primary cabins (i.e. transformer stations) connected to the transmission grid and from medium low voltage (below 1 kV, generally 400 V) starting from secondary cabins, and then reaching the user at the applicable voltage value (230 V for residential and 400 V for industrial). This complex system is regulated by a central brain (Terna's National Control Centre). The TSO in Italy is Terna S.p.a., which is responsible for dispatching, i.e. the management of energy flows on the grid. The explication of the dispatching function requires the monitoring of electricity flows and the application of the provisions that are needed for the coordinated operation of the system's components, i.e. the production plants, the transmission grid and auxiliary services. The DSO, on the other hand, is in charge of the transport and delivery of electricity to the final customer, through the medium and low voltage distribution networks through the city networks. The DSO manages the distribution network and the meters of which he is the owner; he also takes care of the consumption reading, which he communicates later for the calculation of the energy bills. At the end of the supply chain, there is the energy retailer (or sales company) that is responsible for the retail sale of energy to the end customer, buying energy from the electricity market or directly from producers and managing the commercial and administrative aspects related to the supply of energy.

The electricity market as it is organized today is relatively recent and constantly evolving. The electricity supply chain includes all the stages that are necessary for energy to be conveyed correctly, from production to consumption. Figure 2.1 shows the *5 main stages* and the voltage level for each stage.

In general, the production companies take care of the generation phase by transforming the sources in nature (renewable or fossil) into electricity. In wholesale, the production companies sell big quantities of energy to the companies responsible for retail sales. The energy produced and sold wholesale, is transported on the transmission grid, from the production companies to local distributors, on the national high-voltage grid. Subsequently, the distribution companies deal with the transformation of electricity from high to medium/low voltage and the distribution to end customers (houses, companies, public administration, ...). Finally, the retail companies manage the relationship with the final customer. In fact, they purchase electricity from the power exchange or directly from producers and manage all commercial and administrative aspects related to the supply of electricity.

In past times, as energy was a national public monopoly, the price of energy was imposed at the national level and therefore there was no market for electricity. The turning point in the electricity business and the consequent birth of a real electricity market took place following the Legislative Decree 79/99, known as *Bersani Decree*: the objective of the decree, in line with the indications of the European Union, was to promote, according to the criteria of neutrality, transparency and objectivity, competition in the activities of production and sale of electricity, while ensuring adequate availability of dispatching services according to criteria of economic efficiency. After the entry into force of the decree, the state-owned company ENEL (Ente Nazionale per l'Energia Elettrica), whose objective was to manage the entire electricity supply chain, was divided into several companies, each with a specific task related to the production, transmission, distribution and sale of energy. In 2004, the *electricity market* became operational and since July 1st, 2007, it started to be completely liberalized.

While pursuing an economic logic, the electricity market is conditioned by certain technical constraints which must be taken into account to ensure safe and reliable operation of the power grid. The main constraint is the electricity infrastructure which, by its nature, is shared by all market players and has structural limits linked to the capacity, namely the amount of energy that can be carried. Secondly, it is necessary to keep voltage and frequency within predefined limits by constantly balancing energy demand and production. Precisely for this reason, the electricity market is divided into several sessions in which the volumes of energy are sold and purchased to meet both economic and technical requirements.

In an ideal condition, the market should be *nodal* and *instantaneous*: the term *nodal* refers to a market whose reference is the single node of the network while the term *instantaneous* refers to the possibility of balancing instant by instant the energy exchanged at each node. It is easy to understand how this condition is absolutely unreachable in the operation of a large real network. This is the reason why it was necessary to introduce new standards that would simplify the management of the network both from a time and spatial point of view. Regarding time, the standard market unit of measurement is generally *one hour*: a longer time span could lead to problems in maintaining the electricity balance, a shorter one would make it too complicated to conduct market operations. Instead, the space is managed with a division of the national territory into *market areas*.

Figure 2.2 shows said division and allows to distinguish 6 zones:

- **North** (Aosta Valley, Piedmont, Liguria, Lombardy, Trentino Alto Adige, Veneto, Friuli-Venezia Giulia, Emilia-Romagna).
- **Central North** (Tuscany, Umbria, Marches).
- **Central South** (Lazio, Abruzzo, Campania).
- **South** (Molise, Apulia, Basilicata, Calabria).
- **Sicily**.
- **Sardinia**.

Fig. 2.2 Italian electricity market zones

If there are relevant nodes within a zone, it is possible for them to conduct an ad hoc *nodal market*. The model of nodal market employs a single centralized auction to clear jointly short-term energy and transmission markets. In this case, the auctioneer chooses the lowest cost bids that balance energy supply and demand in each node considering the transmission constraints and the physical laws governing the network power flows. In the absence of differences among nodes, the market is configured as *zonal market*. The greater the extension of a zone, the greater the network constraints and therefore more complex will be the operations of adjustment of power flows to ensure electricity balance. The price of electricity is defined zone by zone in the power exchange: on the basis of zone prices is calculated the *Prezzo Unico Nazionale* (PUN), defined as the weighted average of the zonal prices of the *Day Ahead Market* (Mercato del Giorno Prima, MGP) in relation to total purchases. The value of the PUN is therefore not fixed but varies in relation to supply and demand and tends to change from hour to hour. In general, the PUN is high at times when it is more complex and expensive to produce energy while it tends to fall at times when energy production exceeds demand. Trading operations are conducted in two types of markets: *bilateral markets* and *exchange markets*. In bilateral markets supply and demand meet independently without the help of predefined platforms, unlike exchange markets where transactions are settled by organized platforms. Exchange markets are the most common in Europe also due to their higher efficiency and lower transaction costs. In general, the remuneration of energy volumes takes place in two different ways: *System Marginal Price* (SMP) or *Pay As Bid* (PAB). The SMP is a uniform price, the accepted bids are valued at the equilibrium price of the system, equal to the value of the last accepted bid (called *marginal offer*). The operators for sale obtain a revenue equal to the price of the marginal offer [€/MWh] multiplied by the total volume of energy sold on the market [MWh]. The PAB is a discriminatory price, the accepted bids are valued at the price offered for each of them. In general, the most used mode of remuneration is the SMP (i.e. MGP), which guarantees greater transparency, lower costs for the system and greater access to the market for new operators. For some specific markets (i.e. *dispatching services market*), the remuneration of the accepted bids is done by PAB mechanism.

2.1.1 Phases of the Electricity Market

The electricity market in Italy is divided into *Spot Electricity Market* (Mercato a Pronti, MPE) and *Forward Electricity Market* (Mercato a Termine, MTE). In turn, the spot electricity market is divided into the *Day Ahead Market* (DAM), the *Infraday market* (Mercato Infragiornaliero, MI), the *Daily Products Market* (Mercato dei Prodotti Giornalieri, MPEG) and the *Dispatching Services Market* (Mercato dei Servizi di Dispacciamento, MSD), see Fig. 2.3.

On the **MGP** are negotiated the hourly schedules for supplied and consumed electricity. Producers submit offers in which they indicate the amount of energy that they will feed in and the minimum price at which they are willing to sell, buyers submit offers in which they indicate the amount of energy that they will consume and the maximum price at which they are willing to buy. With reference to the day of delivery, the MGP session opens at 8.00 a.m. on the ninth day before delivery and closes at 12.00 noon the day before (hence the name of the market). The market outcome is defined according to an SMP auction mechanism and is communicated after the closing of the market session. Accepted bids are valued at the hourly zonal price, while accepted bids are valued at the PUN.

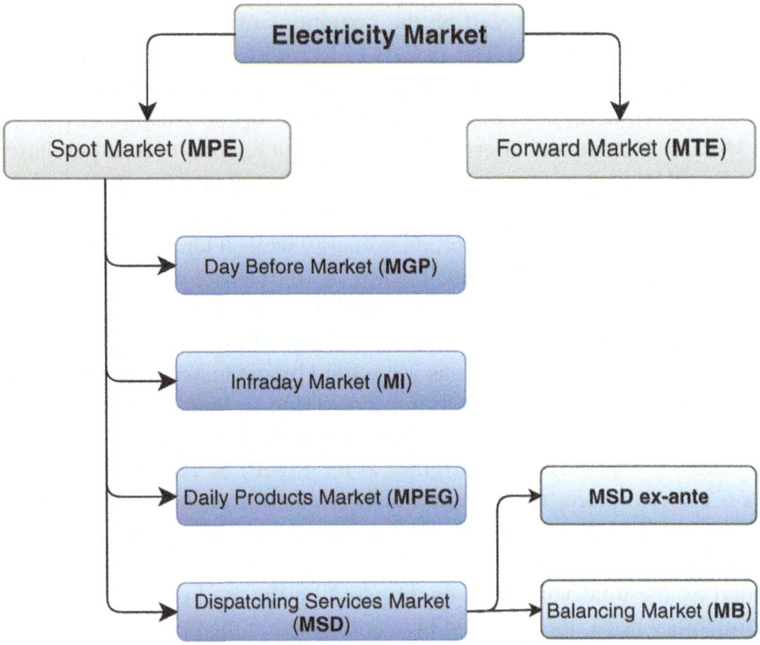

Fig. 2.3 Phases of the Italian electricity market

The **MI** allows operators to make changes to the injection and consumption programs defined in the MGP through additional offers to buy or sell. The MGP is divided into 7 sessions and opens at 12.55 p.m. the day before delivery. The last session, which determines the closing of this market phase, opens at 5.30 p.m. of the day before delivery and closes at 3.45 p.m. of the day of delivery. Also in this case, as for the MGP, the mechanism of remuneration is of type SMP but, to difference of the MGP, both purchase and sale are valued to the hourly zonal price.

In the **MPEG** the trading of daily products with energy delivery obligation takes place. On the MPEG are negotiable daily products with:

- *Unit price differences*, for which the price indicated in the formulation of the offers is the expression of the differential, compared to the PUN, at which the operators are willing to negotiate such products.
- *Full unit price*, for which the price indicated in the formulation of the offers is the expression of the unit value of exchange of the energy object of the contracts.

The delivery profiles negotiable on MPEG are *Baseload*, quoted for all calendar days and where the underlying is the energy to be delivered in all relevant periods of the trading day, and *Peakload*, quoted from Monday to Friday and where the underlying is the energy to be delivered in the relevant periods from the ninth to the twentieth of the trading day. MPEG sessions take place on weekdays from the second day before delivery until 5.00 p.m. on the first working day before delivery.

It follows that on Fridays, for example, both products with Saturday delivery and products with Sunday, Monday and Tuesday delivery will be negotiate

In order to guarantee the correct and safe functionality of the electrical system, the balance of injection and consumption must be maintained moment by moment in order to guarantee the continuity and safety of the supply of the service. For this purpose, the dispatching service has been created, defined as "*the activity aimed at providing arrangements for the coordinated use and operation of production facilities, transmission grid and ancillary services.*" (Legislative Decree 79/99). The dispatching service is provided in Italy by Terna S.P.A. (TSO) according to conditions defined by the Authority. Through the Dispatching Services Market, Terna purchases the resources necessary for the management and control of the system. In this particular market, accepted offers are remunerated at the price presented according to the PAB methodology. The **MSD** is divided into two distinct phases: the programming phase, or ex-ante MSD, and the Balancing Market (MB). The ex-ante MSD is divided into six programming sub-phases. The bidding session on the ex-ante MSD is unique and opens at 12.55 p.m. on the day before the day of delivery and closes at 5.30 p.m. on the day before the day of delivery. On the ex-ante MSD, Terna accepts offers to buy and sell energy for the resolution of residual congestion and the constitution of reserve margins. The MB is divided into 6 sessions. For the first session of the MB, the valid offers presented by the operators in the previous session of the ex-ante MSD are considered. For the other MB sessions, the relevant bidding sessions all open at 10.30 p.m. on the day before the day of delivery and close 1.5 h before the first hour which can be negotiated in each session. On the MB, Terna accepts offers to buy and sell energy in order to carry out the secondary regulation service and maintain the balance, in real time, between energy input and withdrawal on the grid. The balancing resources negotiated on the MSD are found by relevant production units enabled (users of the dispatching of enabled units are obliged to submit bids). Through pilot projects, at the moment they can also submit bids other than the classic relevant units identified by the grid code as long as they are enabled according to the regulations of these projects. Table 2.1 shows a summary diagram of the Italian electricity market in which is possible to distinguish phases, aims of the exchanges, operators and remuneration mechanisms.

Table 2.1 Italian electricity market organisation scheme

	MGP	MI	MSD	
Exchanged resource	Energy	Energy	Energy for grid congestion resolution	Energy for real-time balancing
Unit allowed to participate	All input and output bidding points		All input and output points enabled for dispatching services	
Operators allowed to participate	Market operators	Market operators	Dispatching users	Dispatching users
Remuneration	SMP	SMP	PAB	PAB

2.2 Actors of the Electricity Market

2.2.1 *Prosumers*

The term *prosumer* is the combination of the terms *producer* and *consumer* and indicates an end user connected to the distribution network (medium and low voltage) that consumes and produces energy. An example are energy consumers, private individuals or companies, who locally install generator systems to cover part of the energy consumed and feed any surplus of energy produced into the grid. The installations mainly concern photovoltaic systems (small and medium size) and micro-wind systems (medium size). These are increasingly associated with storage systems, usually electrochemical, for better production management. The technological experience reached and the reduction of the production cost of the batteries guarantees an optimal solution to the problems related to the intermittence of the source and to the peaks of production. This does not mean, however, that the prosumer must or wants to be *offgrid*, i.e. detached from the grid: in fact, all users are normally connected to the grid, either active or passive. Sometimes storage systems can be installed even in the absence of a production plant in order to accumulate the energy coming from the grid in periods of low load, and at lower prices, and use it in peak periods thus obtaining a saving. The current Italian technical norms (CEI 0-16 and CEI 0-21) identify a storage system as a generator, so a consumer who installs a storage unit at home automatically becomes an active user and therefore a prosumer. It is therefore interesting to evaluate the network/prosumer interactions and those between prosumers themselves: these interactions can be well mapped to blockchain applications in the electricity market, which guarantee data tracking in a safe and reliable way. End users, whether active or passive, no longer represent the last stage of energy delivery but can become part of the grid regulation processes through, for example, participation in DR programs. Participation in such programs is subject to the signing of a supply contract and the regulatory contributions are not paid out individually but as an aggregate. From this point of view, a prosumer can be a residential user or a large company, it can have an adjustable energy structure; in any case, the interface to the service is managed by a third party responsible for the process defined *aggregator*.

2.2.2 *Aggregator*

The participation of the prosumers in the DR service takes place through their aggregation in virtual units, which are constituted and managed by a subject known as "aggregator". Following the proposal for European energy efficiency Directive recast and the Directive 2019/944/EU on the internal electricity market, the "aggregation" is defined as: "a function performed by a natural or legal person who combines multiple customer loads or generated electricity for sale, purchase or auction in any electricity market". While "independent aggregator" is "a market participant engaged in aggregation who is not affiliated to the customer's supplier". The aggregator must thus respond to the modulation orders given by the grid operator (TSO or DSO), so as to prevent any imbalance of load and generation in the power grid arising at certain times of the day when the energy supply does not meet the demand [1], see Fig. 3.5. The aggregator meets these orders by in turn asking, in compliance with the constraints and requirements for the aggregate resources, the modulation of the consumers' load that make up the virtual load unit. The term "aggregator" originally was referred to aggregate loads management, more recently, this entity can also aggregate generation units thus giving rise to a "Virtual Power Plant", i.e. an entity which is responsible for managing multiple generators [2]. Aggregators are therefore third party companies that initiate contracts with individual end users (residential, civil, industrial) in order to create an aggregate operating as a Demand Side Response Provider (DSR Provider). Through an appropriate combination of actions on the load profile of individual users, the DSR Provider provides the network with an active power regulation service [3]. The actions generally relate to changes in the power consumption and/or supply of the users participating in the programs and are carried out in response to the needs of the network. The effect of these changes is to support the grid in the presence of contingencies such as peaks in demand, grid congestion, overloads or other failures, avoiding or otherwise reducing the risk of system failure and ensuring more economical and efficient operation. The task of the aggregator is therefore to supply "flexibility" to the system receiving an economic remuneration tied to the entity of the offered service. The responsibility for the entire process lies with the aggregator, who must eliminate the uncertainty linked to the individual user behaviour and guarantee the provision of the service. An example of aggregator operating in the Italian territory is Enel X: since 2017, the year in which the Regulatory Authority for Energy, Network and Environment (hereinafter referred to as ARERA or the *Authority*) allowed the participation of the power demand in the Dispatching Service Markets (ARERA 300/17), Enel X has played the role of aggregator distributing the modulation order among the customers in its portfolio, who are available to reduce or increase consumption and modulate load and/or generation. As an aggregator, therefore, Enel X generates value for customers, identifying, enabling and rewarding the flexibility of the load, and, at the same time, generates a value for the network quickly providing a certain dispatching capacity. In the UVAC (Unità virtuali di aggregazione di Cario or Virtual Unit of Aggregated Consumptions) experimental project (ARERA 372/17), Enel X was the main aggregator with the award of over 110 MW of installed capacity across over 30 sites [4]. As regards the UVAM (Unità Virtuali di agrenazione mista or Virtual Units of Mixed Aggregation) project (ARERA 422/18), 25 different aggregators have already been allocated capacity: Enel X holds the highest share among all the entities involved with 41% of the market share (694 MW) [5]. In January 2020, in Italy, the first project for aggregating residential storage units to offer balancing services to the grid was started. The project was implemented in the provinces of Brescia, Bergamo and Mantova. The provision of balancing services was in the past reserved only to large production plants or industrial loads, now residential users are also able to participate in demand management programs in the form of aggregates.

2.2.3 Energy Markets Manager

The Manager of Energy Markets (Gestore dei Mercati Energetici or Manager of Energy Markets, GME) is the company responsible for the organization and management of the Italian electricity market. Established by the Manager of Energy Services (Gestore dei Servizi Energetici or Manager of energy Services, GSE), it is fully owned by the Ministry of Economy and Finance and carries out its activities according to the indications of the Ministry of Economic Development and the Regulatory Authority for Energy networks and Environment (Autorità di Regolazione per l'energia Reti e Ambiente, ARERA). The aim of the GME is to promote competition among producers by ensuring the availability of an adequate level of power reserve. In the Day Ahead Market and Infraday Market, producers sell energy and buyers buy it wholesale: on both markets, the GME acts as a counterpart in trading. It also manages the Dispatching Services Market where the counterpart is TERNA. The capacities related to DR programs by aggregates are currently allocated according to the regulations of the pilot projects: in the future, given the purpose of the programs, the offers will be presented in the MSD. The GME plays a specific role in the market monitoring for the Italian Regulatory Authority, originally resulting from deliberation n.115/2008 of the ARERA, containing specific provisions for the monitoring of the electricity market in Italy, and later reinforced by the provisions of Regulation (EU) n. 1227/2011 on the integrity and transparency of the wholesale energy market [6].

2.2.4 Balance Responsible Party and Balance Service Provider

The *Balance Responsible Party* (**BRP**) is a market operator or its representative, responsible for its own imbalances in the electricity market, as defined in [7]. A BRP (or dispatching user) can be an end-customer or producer or its intermediary (wholesaler) who signs the dispatching contract with the System Operator and who is responsible for the execution of the feed-in and withdrawal programs negotiated on the wholesale markets. The term *Balance Service Provider* (**BSP**) refers to the final customer or producer or its intermediary (wholesaler) who is responsible for providing the dispatching services to the System Operator, which are necessary to maintain a real-time balance between supply and demand for electricity. The figure of the BSP coincides with the role of the aggregator and may or may not coincide with the dispatching user [8]. The BSPs provide the TSO with the resources for the *ancillary services*, i.e. services necessary to guarantee the security of the entire electrical system. Providing ancillary services essentially means modifying one's feed-in or withdrawal programs in order to respond to the TSO's requests and offering a regulatory contribution to the network. In general, BRP injection and withdrawal programs can be formed in the Day Ahead Market, MGP, and/or in the Infraday Market (MI) and can subsequently be modified by the TSO by accepting offers submitted by BSPs in the Dispatch Services Market (MSD) to take network constraints into account. It is also possible that a BRP has programs that are independent of MGP and MI, in this case, dispatching takes place entirely on the MSD. The objective of balancing is to ensure a continuous balance between inputs and withdrawals in order to maintain the performance requirements related to system frequency (EU Regulation 2017/1485). A general review of the dispatching discipline is underway in relation to the new market environment and the increased flexibility requirements of the system. The concept of "aggregation" was conceived as a response to these needs and allows, through figures such as BRP and BSP, to supply the system with reserves of flexibility previously inhibited by norms.

2.2.5 Transmission System Operator

The *Transmission System Operator* (TSO) is a company responsible for the control and management of the transmission network. The main task of the TSO is to transport electricity over long distances through the High-Voltage grid (from 150 to 380 kV in Europe) that connects large production centers (thermal power plants, nuclear power plants, large renewable energy sources) to industrial users' substation or distribution networks. Due to the high costs for the installation and maintenance of electrical infrastructures, the TSO generally operates as a natural monopoly and is subject to state regulations. In Italy, Terna plays the role of TSO. The management of the grid includes the monitoring of power flows and voltage control in all nodes of the system: the TSO guarantees the stability and reliability of the power system by balancing demand and production of electricity in real time.

The dispatching service is totally managed by Terna: through dispatching, the units that provide the supply of services needed to guarantee the security of the national electricity system are identified. Moreover, the regulation of the economic items deriving from the execution of the contracts and from the actual imbalances is completed [8]. The dispatching service is supplied on the basis of contracts signed with BRP, (single units), or with BSP (eventually aggregated units). Through the MSD, the TSO attains the necessary resources for the supply of ancillary services by selecting the most economical offers formulated by the BSPs. In case of particularly serious and unexpected contingencies, the supply of ancillary services can happen also through forced actions, even if not remunerated.

Participation in MSDs implies that the service delivery units are enabled for the purpose. The network code provides for the modalities of enabling units considered traditional: in response to the evolution of the energy context and its needs, in recent years, several pilot projects have been issued with the aim of assessing the impact on the system of units that previously were not considered suitable to provide ancillary services, such as power plants from renewable sources or small consumption and/or production units ($P < 10$ MW). For these projects, the Authority (ARERA) indicated Terna as the body for the definition of the new technical performance requirements of the relevant production units and the new Enabled Virtual Units. Terna's task was to identify, initiate and manage the projects, drawing the necessary conclusions regarding their impact on the system. Further detail are given in Sect. 2.5.1.

The TSO plays a fundamental role in the management of the electricity system: it not only represents the main actor in all the crucial phases of the system's operation, but also operates, under the Authority's instructions, a constant action of research and innovation of the grid, the results of which will be the basis of future regulations.

2.2.6 Distribution System Operator

The *Distribution System Operator* (DSO) is a company responsible for the management of low and medium voltage distribution networks that it sometimes owns. High voltage transmission grids have the task of transmitting large amounts of energy over long distances (national extension): in the vicinity of end users, electricity undergoes a voltage transformation and is distributed through the distribution network (regional extension). The distributor's tasks include making new connections, solving grid faults and reading energy consumption via meters (of which it is the owner). Despite distributing it, the distributor does not physically supply the energy: there is an important difference between the company that distributes the energy and the one that sells it. The Authority requires two different entities to distribute and sell electricity and, in case a company takes care of both operations, the latter is obliged to create two different companies (one for sale and one for distribution). The assignment of the distributor is made by geographical area through assignment by the ARERA: this implies that the consumer cannot choose the distributor but only the energy supplier. The most important distributor in Italy is E-Distribuzione, which distributes energy to more than 90% of consumers on the national territory.

In the context of large penetration from unpredictable sources at the distribution level, the main task of DSOs remains to provide a secure supply of electricity while ensuring an acceptable quality of service, but in order to pursue these objectives they must manage an active network increasingly complex and articulated. As far as network balancing is concerned, the supply of ancillary services by load aggregates (connected to medium and low voltage networks) implies an evolution of the distributor's role: the distribution companies will have to facilitate the supply of global ancillary services provided by BSPs and, in the near future, will also acquire the role of buyer of resources for local ancillary services, i.e. services necessary for the safe operation of distribution networks or portions of them only. The actual effectiveness of local ancillary services will have to be tested and validated by special pilot projects before the practice becomes standard and is included in the network codes [8].

2.2.7 *Energy Retailer*

Since the issuance of the Bersani Decree, several figures in the energy business were born, figures that were previously unnecessary because the market was a state monopoly. A new player, appeared from the liberalization of the market, is the *Energy Retailer* (also called *supplier*). Since January 1st, 2007, all customers have the possibility to choose their supplier on the free market. In the transitional period that has as its objective the total liberalization of the electricity market, scheduled for 2022 [9], the Authority has established a service defined of *Maggior Tutela*, a service in which the contractual and economic conditions of the supply are defined by the ARERA itself. The customers are therefore divided into customers of highly protected customers (Maggior Tutela) and customers of the free market. Only for the latter, the suppliers manage all commercial and administrative aspects of the electricity supply. In Italy there are currently over 40 Energy Retailers. Among them, the majority share of sales to the end market belongs to *Enel Energia*, with about 35% of the market share [10].

2.3 The Blockchain for the Electricity Market in Italy and Europe

Heavy computational effort can come after the scale-up of blockchain solutions for different energy markets. For this reason it is important to understand how, in the current organization of the electricity market, the blockchain technology could be used. The electricity market in Italy, such as in most European countries, has largely changed in recent times due to the transition to a low-carbon economy. Electricity is now traded at the wholesale level or at the retail level. Trading takes place between those who offer electricity (generators) and those who demand electricity at the wholesale level (traders), and between suppliers and end-users at the retail level. Recently, however, are appearing new forms of energy trading. The latter are called P2P markets. In this case, electricity would be traded between peers, namely end-users. To better understand how blockchain can support the implementation of the different electricity markets, it is important to better detail the electricity markets organization. Considering the type of transactions, the electricity market can further be divided into regulated and non-regulated markets. The former comprises spot and forward markets, while the latter in most cases includes the so called over the counter, OTC, markets, which typically are forward markets based on bilateral contracts. A large part of the activity inside non regulated markets is devoted to barely financial transactions. The main issue with these transactions is the limited access from operators. Such problem could be addressed through the blockchain technology, as it would allow access to all actors of the market. In the regulated markets in Italy, operators must be recognized by the GME, which is a trusted third party providing the permit of take part to the market , provided some conditions apply. In this case, the scalability issue can be considered not as a primary requirement, as the transactions are future deliveries. Both spot and forward markets are wholesale electricity markets. The main difference between the two markets is the way in which the deal is implemented. In the case of spot market, the deal is arranged into market sessions, with a given reference period for delivery, and in which each party offers its capacity and price. In the Day-Ahead Market (MGP), the deal is constrained by market sessions in which the price setting is hourly; in the forward market instead the deal is continuously carried out. In this case, price, volume and the delivery period can be negotiated (futures). In the two cases, the deal is carried out on digital platforms (IPEX, EEX …) that collect offers and implement the clearing algorithm. In the so called forward markets, that are supported by bilateral trading, both supplier and consumer agree on a set of conditions that are part of a trade contract and they interact directly with each other. In this latter case, trading can keep the market price released by the power exchange as reference price. In organized bilateral trading, generation and demand bids are submitted to a platform by those entities which take part to the market. The market platform is cleared continuously. One of the market players can accept the bid of another market player, resulting in different prices for each trade. The "exchange markets" (spot and forward) have as counterpart in Italy the GME. In this case, bilateral deals occur between two parties and thus there is no central counterpart, but the physical delivery must always be accepted by the market manager (GME in Italy). For this reason, the outcomes of non regulated platforms for OTC must be notified to GME and then to the TSO. A generic transaction on the futures market requires some information: the identification of the parties in the contract, the amount of energy exchanged, whether it is offered or purchased, the profile with which the generation is offered (baseload, peak, offpeak), and the duration (month, three-months, year). The use of the blockchain technology to implement wholesale electricity markets, and specifically, regulated and non regulated bilateral (OTC) markets, may reduce the counterpart risk and may increase transparency while setting aside the privacy-preserving feature. As an example, the Enerchain project [11] is developing a blockchain-based P2P trading platform to more transparently complement, or replace, the wholesale electricity market. The retail market requires a wholesale buyer that can then sell to individual customers. This is, for example, the case of aggregators. In this latter case, blockchain technology is able to improve transparency towards end-users and between the TSO and the aggregator. The European Directive on common rules for the electricity markets claims that "integrated short-term markets are missing". And recognizes that short-term electricity markets would allow "trading RES-E across borders" and they are the "key for the successful integration of RES-E into the market". Furthermore, the same Directive underlines that "the advent of digitalization, internet-based metering and trading solutions enable industry, businesses and even households to generate and store electricity, as well as participate in electricity markets via Demand Response solutions.". Indeed, the electricity markets of the future will be characterized by more variable and decentralized electricity production, an increased interdependence between Member States across the borders and new technological solutions for end-users to reduce their bills and actively take part to electricity markets through Demand Response, selfconsumption or storage [7]. The cited Directive also supports local energy communities as a means to foster renewable energy penetration. Nonetheless, it seems that the so-called P2P markets are still in an experimental phase. The paper in [12] makes a large review of P2P markets. The work outlines 3 different kinds of P2P markets: full P2P market; community-based market; hybrid P2P market. In the full P2P market, agents (i.e. producers, consumers and prosumers) can interact even without third parties (i.e. retailer or market operator). In this case, a couple of agents may agree on a certain amount of energy exchanged at a given price without central supervision. The agents should thus be able to understand if the transaction is technically feasible and to what extent it may influence other nodes in the power network. The second type of P2P market, the community-based market, appears more structured. This market requires the presence of a third party, called community manager, to regulate transactions among agents inside the community. This party also governs the interaction of the community with the main grid. The last kind of P2P market, the hybrid one, is a merger of the two previously described models, ending up with different layers for trading energy. In each layer, communities and single agents can interact directly. This latter model is more organized than a P2P market but not as structured as a community-based market. As it can be understood from this description, P2P markets are a natural experimental arena for blockchain technology and distributed ledgers. What seems evident is that while wholesale markets (forward) do not consider technical feasibility of transactions that are then fixed in other shorter-term markets or even after the time of delivery, P2P markets may need a strong technical management while transactions are carried out, as there no possibility to fix possible technical problems arising as a consequence. In conclusion, the request raised by the Directive about reinforcing the short-term markets appears more practical, although the feasibility of the use of blockchain solutions considering some time for validation, consensus and conflicts resolution must be always taken into consideration. In principle, addressing a full P2P market would require the non-existence of the power network considered as place where possible imbalances can be compensated, leaving the full operational control to decentralized agents. From this point of view, the forecast of loads and generation, as well as local availability of storage, would enable the implementation of full P2P electricity exchanges, making use of the blockchain technology. In conclusion, as pointed out by the Directive, the new electricity market in Europe should be, as much as possible, "integrated" and efficient. As concerns integration, the way the price is formed should be shared and agreed upon among all actors in Europe. This could be easily carried out by means of a clear business integration tool supported by the same protocol (i.e. blockchain technology). As far as efficiency is concerned, the values of all assets at any time should give back all obtainable information. In a marketplace that operates efficiently, indeed, suppliers and buyers together can adjust the price of products in real-time considering the new information available. Both integration and efficiency can thus be acquired using the blockchain technology as an facilitator of B2B and B2C integration and transparency in data exchange. Table 2.2 summarizes the main features to be considered in electricity markets implementation.

Table 2.2 Features involved in the electricity markets implementation

	Spot markets	Bilateral (regulated and OTC)	Blockchain and P2P
Control over energy transaction	Centralised	Decentralized	Decentralized
Number of participants	Many	2	Those taking part in the BC community
Conditions for entering the market	Yes	Yes	No
Participants transparency	None	Between the two participants	Verified by the authentication mechanism
Timing	Mostly day or weekly	Mostly hourly	15' to hourly
Trading	Through the market platform	Directly or through a trader	Directly
Price setting	Merit order or Demand/Offer agreement or "pay as bid" when needed	Influenced by exchange markets	Influenced by exchange markets
Price transparency	To all	To those who take part in the transaction	To those taking part in the BC community
Counterparty risks	Minimum	High	Minimum
Need for balancing	Strong	Strong	Limited
Contract and products	Standardized and subjected to precise rules	standardized	Not standardized

2.4 Business Models and New Energy Services

If we think about the traditional energy transaction model we can say that it is a centralized model, where all energy transactions are performed through trusted entities (GSE, TERNA, Distributors and Suppliers) that manage the whole process from production to delivery to the end user. As it happens for the financial transactions, in which the bank acts as an intermediary to process the electronic payments. This system is inherently weak as it is based on a model in which there is one trusted third party that acts towards two unknown users. This results in an intermediation cost that surges the transaction cost, setting a cap on the minimum size of viable transactions and preventing the implementation of small transactions. In fact, in the electricity market, the residential user does not have the possibility to buy directly from the producer and bargain on the purchase price, but buys from a supplier at the PUN, which has purchased energy wholesale. The same concept applies to small producers from renewable sources. The energy produced is injected into the grid and remunerated at a price set by the authority. This energy then reaches the end user through the TSO, DSO and energy retailers and there is no direct connection between producer and consumer. The use of blockchain in the financial world, to date, has made it possible to overcome this kind of problems, just think of Bitcoin or Ethereum, which allow the exchange of currency between two unknown users without the need for a third party to act as guarantor for both. The direct consequence is the reduction of the transaction cost, since the exchange takes place between the users involved in the transaction without the need for an intermediary, thus also allowing the possibility to perform small transactions. This powerful new technology therefore allows transactions of a valuable asset directly between end users without the need for an intermediary to act as guarantor. And what would happen if the asset was energy? Every user of the network would have the possibility to interface directly with the producers and vice versa, with a consequent reduction in transaction costs and therefore energy. From a centralized transaction model, we move on to a distributed model, in which all network users can buy or sell energy from whoever they want or from whoever best meets their needs, generating new business models of the electricity market and new management logics of network services such as ancillary services.

2.4.1 Energy Business Models Enabled by the Blockchain

The increasing use of small power generators, mainly solar photovoltaic (PV), connected to the distribution network and the increasing electrification of transport and buildings are more and more leading to the digitalisation of the electrical system. All new assets on the offer and demand side are adding complexity to the energy sector, making monitoring, management and control crucial to the success of energy transformation. The energy market has for years been open to prosumer, the small renewable energy producer that does not consume everything it produces. But the market model, as described above, is still centralized, with inefficiencies and high costs. Digital technologies such as blockchain can support renewable energy penetration by improving monitoring, operation, maintenance and control and generating new market mechanisms. Thanks to the blockchain in fact, new prospects have been opened for the creation and diffusion of new markets for energy where people and companies with an "individual production" can directly sell their excess production capacity to the market or buy it in case of need, creating new scenarios that can also bring important benefits in terms of balancing networks. The use of this new technology in the energy sector has contributed to the evolution of new business models,with the focus on local P2P and wholesale electricity trading, as well as innovative means of financing projects in developing countries. A key tool provided by the blockchain that enables further development of initiatives in the energy sector are SCs. Thanks to SCs, it is possible to automate the rules and sanctions relating to an agreement and automatically enforce obligations, removing the need for an intermediary. The implementation of SCs performed on blockchain networks can help modernize electricity grids, improving the use of renewable energy, operation and management of the grid, allowing especially the development of new market logic. SCs could also be used to automatically buy and sell energy for end-users or companies based on real-time price signals. End-users become active participants in the market, able to buy and sell their electricity without the need for trusted authority or intermediary. The use of blockchain technology in conjunction with SCs allows the development of efficient, effective and affordable services to help transform the energy sector. In addition, blockckhian, as a tool for data management, facilitates coordination between different actors because everything is shared, transparent and immutable.

According to [13], the new business models enabled by blockchain in the energy sector are many, but those that are more feasible and that promise value creation throughout the energy sector are related to:

- **Certification of origin systems for renewable energy markets**: applications that document the origin of renewable energy production, release of certificates for each unit of renewable energy produced and tracking transfers of ownership between market players for their statements on green energy and compliance requirements.
- **Demand Response programs**: applications that perform real-time aggregation, measurement, tracing and verification, settlement and trading associated with participation in a given demand response event without the need for an aggregator to interface customers and DSOs, with the potential to involve domestic users in the programs.
- **Electric vehicle charging networks**: applications for the management of customers, vehicles and charging infrastructure by means of cryptographic identities to provide services to the electricity grid.
- **Transactive energy systems**: market projects where electricity grids are balanced and controlled through the execution of P2P energy exchanges between end users. Thus allowing the purchase and sale of energy between neighbors and encouraging the installation of domestic production systems.
- **Wholesale energy trading**: market projects where the blochckain is used to reduce the counterparty risk, improve transparency and to preserve privacy in the wholesale market.
- **Grid management**: applications for the management and tracking of the supply chain and maintenance of power grid.
- **Metering and billing**: applications for tracking and certification of user consumption.

For all the considered applications, there are some specific and general challenges concerning the use of the blockchain technology [14, 15] that are worth mentioning and that are listed in Table 2.3.

2.4.2 The Blockchain Technology for Ancillary Services Provision

Else than P2P electricity transactions, V2G or DR programs, a further application of the blockchain technology in the electricity field could be the network management operations referred to voltage and frequency regulation, namely the so-called ancillary services. As a result of decentralization and of RES penetration, the number of

Table 2.3 Specific and general challenges

Application	Specific challenges	General challenges
Peer to Peer and decentralized energy trading	Implementation of efficient market algorithms as smart contracts; Balancing demand to supply and need for machine learning algorithms; Interoperability of current smartmeters with blockchain platform	– Compatibility with low power nodes – Regulatory challenges – Costs of integration – Latency – Privacy and GDPR compliance – Scalability (management of large amount of data) – Bandwidth – Power consumption
Wholesale energy trading	Involvement of all market actors	
Energy trading support for small generators and end-consumers	Interoperability of current smart meters with blockchain platform; Compatibility with existing market structure	
Metering and Billing	Interoperability of current smart meterswith blockchainplatform; Metering data verificationfrom multiple actors for consensus	
Vehicle to Grid	Interoperability with current existing V2G control technology	
Demand Response for Primary/Secondary/Tertiary regulation	Communication of current smart meters with blockchain platform; Metering data verification from multiple actors for consensus; Latency; Bandwidth compatibility	
Green certificates and carbon trading	Communication of current smart meters with blockchain platform; Metering data verification from multiple actors for consensus	
Grid management	Higher throughput and transaction speeds that would allow real-time verification; Interoperability with current existing metering technology; Generation of massive datasets	
Power Purchase Agreements	Interoperability of current smart meters with blockchain platform; Metering data verification from multiple actors for consensus	

transactions needed to guarantee ancillary services provision will largely increase. For P2P trading, the goal of enabling technologies such as blockchain is to support and make easier electricity trading between users, reducing third parties intervention for ancillary services provision. These third parties, i.e.: system operators and balancing responsible parties, BRPs, could directly trade this provision with prosumers. Indeed, as the latter have to be notified about the demand of real or reactive power injections into the network, this service provision from end-users could lead to a direct trade between end-users and system operators or BRPs. For such implementation, again, a transparent and secure blockchain, to which all users can take part, is needed but with a verification and consensus protocol controlled by a pre-defined set of nodes, named verification nodes. In this, such as in other cases, the algorithm for deciding which node should provide how much power and at what time is an "optimal power flow". The main prospect of blockchain technology, in this scenario, is that it can provide a secure technical basis for the implementation of different markets. The use of blockchain technology supports the direct provision of ancillary services and, at the same time, reduces barriers to market entry for smaller parties connected to the network. As electricity grids are critical infrastructures, they operate based on many different rules to ensure that entities taking part to ancillary services provision respect the contractual agreement they have with network operators [16]. Due to these complex regulations, small prosumers do not provide ancillary services, although in principle they could. By means of blockchain technology, SCs could be used to implement the current regulatory system for ancillary services provision into the digital world. Once the SCs have been defined, they can be employed to smooth market processes for the tracing of ancillary services in an efficient and secure manner. Moreover, these SCs could facilitate market activities for ancillary services provision from all users in the network (from large industrial consumers to residential users). An instance of this approach is the D3A framework created by the Energy Web Foundation, pointing at the integration of decentralized assets including residential loads in local and national markets for energy and ancillary services. WePower and the Estonian TSO Elering or Tennet with the battery storage company Sonnen are now analyzing how SCs could be used to implement functions that help to balance the network. In the view of a future scenario, network management will become more and more digitalized with features that traditional grid operators could not supply. As a result, it could be decided to entrust network operations to a different actor that can manage digitized power systems more productively. This model is currently implemented in transmission grids across United States, where utilities own energy resources and infrastructures, while the Independent System Operators (ISO) manage network operations. Despite the technological barriers are still numerous (i.e.: the management and accounting of assets and the truthfulness of the measurement of the same assets) blockchain can bring a positive contribution to overcome them [17].

2.5 New Energy Regulations

The decarbonisation of the national energy system and the consequent shift to a mix more based on the contribution of RES, wind and photovoltaic, is certainly an important objective from an environmental and economic point of view, as well as being substantially imposed by European directives. But a transformation like this has significant consequences for the electricity system and the electricity market, which in some way must be governed, to avoid problems on the availability of electricity for families and companies. The point is that the increase in the production of energy from RES inevitably reduces the contribution of production from conventional power plants, which in recent years have decreased in number or, in any case, tend to produce less energy. The problem is that the new generation from RES are not programmable, they are unable to generate electricity 24 h a day. The growing share of energy produced by non-programmable RES power plants, especially photovoltaic, is concentrated in specific hours of the day. This leads to the need for more power margin to ensure the safety of the system.

To address these issues, the European Council has adopted two directives of the "Clean Energy Package", presented by the EU Commission in 2016, on the promotion of the use of energy from RES (2018/2001/EU, RED-II) and on increasing energy efficiency (2018/2002/EU), in addition to the Regulation on the governance of the Energy Union (2018/1999/EU) and the Directive on energy efficiency in buildings (2018/844/EU). These directives set a target of production from RES of 32% and an increase in energy efficiency of 32.5% by 2030, while the governance regulation provided for the submission of national energy-climate plans by the end of 2018. The new governance regulation facilitates mutual assistance between Member States in the event of an electricity crisis, so that they can benefit from better preparation and less interruptions during peak periods. To this purpose four measures of the Clean Energy Package have been adopted. Directive No. 2019/944/EU of 5 June 2019 and Regulation No. 2019/943/EU of 5 June 2019 are related to the internal electricity market and aim to make it more flexible and competitive, taking into account the increasing penetration of production from RES. To avoid financing fossil fuels, the regulation provides for an emission limit of 550 g of fossil CO_2 per kWh of electricity: new power plants with higher emissions will not be able to participate in the capacity markets by providing energy in case of need to ensure the security of the electricity system. Old power plants will be allowed to continue, but only under certain conditions and until July 1, 2025. The other two measures, Regulation 2019/941/EU and 2019/942/EU are related respectively to the prevention of blackout risks and cooperation between national energy regulators and complete the Clean Energy Package.

These directives have then been adopted by the various European countries to achieve the objectives set by the EU Commission. Below are described the main directives and applications developed in Italy with the aim of promoting production from RES, increasing energy efficiency, the competitiveness of the electricity market and ensuring the reserve of energy necessary for the correct operation of the power system.

2.5.1 Aggregation of Local Energy Resources

The electrical system must be able to guarantee at all times that the energy required by all consumers is balanced by the energy produced by the power plants. During the last decade, this power margin has always been guaranteed by conventional power plants, but in the last three years the legislator and the ARERA (or the Authority) also in response to the directives of the EU Commission have decided to experiment with an expansion of the entities entitled to contribute to the balancing of power flows on the electricity system. Already in 2015, the Authority started a process of reform of the dispatching service in a manner consistent and coordinated with European norms. The objective is to enable all relevant production units to participate in the MSD, including those supplied by non-programmable sources, relevant consumption units and non-relevant production and consumption units in aggregate form, also through DR programs. In order to achieve this goal, the authority has indicated Terna as the institution for defining the new technical performance requirements of the relevant production units and the new Virtual Enabled Units (Unità Virtuali Abilitate, UVA) and for identifying, starting up and managing projects, obtaining the necessary conclusions regarding their impact on the system. The regulatory starting point is represented by Legislative Decree 102/2014 (today replaced by Legislative Decree 73/2020) which represents the transposition of the European Directive 2012/27/EU on energy efficiency (today replaced by 2018/2002/EU), with the aim of establishing a framework of measures for the promotion and improvement of energy efficiency that contribute to the achievement of the national energy saving target. Following this decree, with Deliberation 300/17/R/eel [18], ARERA has promoted the opening of the MSD to electricity demand, non-programmable renewable sources and storage systems, with experimentation through pilot projects. In particular, the Deliberation distinguishes between:

1. **Consumption enabled virtual units** (Unità Virtuali Abilitate di Consumo, UVAC), characterized by the presence of only consumption units.
2. **Production enabled virtual units** (Unità Virtuali Abilitate di Produzione, UVAP), denoted by the presence of only non-relevant production units, whether programmable or non-programmable, including storage systems.
3. **Relevant production units** (Unità di Produzione Rilevanti, UPR), characterized by the presence of relevant generating units powered by non-programmable renewable sources and relevant generating units that meet the minimum technical licensing requirements for at least one of the dispatching services between: congestion resolution at the planning stage, tertiary reserve resources and balancing resources.
4. **Relevant production units integrated** with storage systems (Unità di Produzione Integrate, UPI), characterized by aggregates consisting of relevant production units (UPR) and storage systems able to guarantee the provision of the primary control service in any operating condition provided by the Network Code.

5. **Mixed enabled virtual units** (Unità Virtuali Abilitate Miste, UVAM), characterized by the presence of both non-relevant production units and consumption units and storage systems.

As the ARERA definition establishes, UVAMs are characterized by the presence of production units (relevant and not relevant), storage systems and consumption units, connected to the grid at any voltage level. More in detail, UVAMs are characterized alternately:

1. by the presence of non-relevant production units (including storage systems, between which may also include the batteries of vehicles that connect to the electricity grid through charging infrastructures equipped with V2G technology), with charging units of relevant production not already obligatorily enabled that share the point of connection to the grid with one or more consumption units, provided that the power input to the connection point is not more than 10 MVA, and consumption units;
2. by the presence of relevant production units that are not already obligatorily enabled and that share the connection point to the grid with consumption units and have a power input at the connection point greater than 10 MVA.

For qualification requirements, UVAMs must be able to make available a capacity of at least 1 MW, to increase and/or decrease. There is no limit in relation to each individual production or consumption unit of which the UVAM is composed. The UVAM systems have been designed with the aim of expanding as much as possible the audience of possible participants. The project, approved with the Deliberation of ARERA 422/2018/R/eel (operative since November 1, 2018) is part of the pilot projects, provided for by the aforementioned Deliberation 300/2017/R/eel, aimed at collecting useful elements for the reform of dispatching and the expansion of dispatching resources. However, there are some pre-requirements for the qualification as UVAM [19], i.e. it must not be qualified to the Capacity Market and must have a modulation threshold of 1MW, the lowest threshold at European level.

As indicated in the regulation, UVAM are enabled to provide services such as congestion resolution, balancing, tertiary reserve, but the plan of Terna and ARERA is to test the participation of UVAM on other types of services, such as secondary frequency/power regulation and to progressively involve also consumers in the tertiary and/or domestic sector.

Already in June 2019, more than 120 UVAM for a total power of 830 MW with more than 25 entities classified as BSP [20] have been enabled to MSD services. For the whole 2020, 991.4 MW have been assigned on the entire national territory, divided into 800 MW for North and Centre-North and 191.4 MW for Centre and South.

Being part of a UVAM can be advantageous because in addition to the ordinary remuneration related to the energy activated (€/MWh) there is also the remuneration of availability of the resource with a fixed fee, calculated in €/MW, unlike for large power plants. It should be considered that industrial operators, who contribute to the development of UVAM, have to bear fixed investment costs to install and set up the equipment necessary to develop the service and annual operating costs. The remuneration of the activated energy alone, which is affected by the volatility of the market price, is therefore generally not sufficient to cover these costs, thus requiring the adoption of a fixed fee to balance the accounts.

One of the most important features of the project is the inclusion of storage systems functional to electric mobility, opening the way to the experimentation of V2G technology. In Italy, V2G technology has been regulated by a Decree of the Ministry of Economic Development of January 30, 2020 (Mise Decree) [21], which establishes the methods of diffusion, the rules for participation in the energy market and the rewards for those who use their electric vehicle to help the national power grid by providing services of tertiary reserve, balancing or frequency and voltage regulation.

The V2G decree establishes that electric vehicles and charging stations can participate in the MSD in aggregate form via UVAM. This is possible because there is no limit in relation to each individual production or consumption unit of which the UVAM is composed. The decree states that in the case of UVAM consisting exclusively of charging infrastructure, the minimum power to be made available can be reduced up to 0.2 MW, increasing and/or decreasing, given the low power of electric vehicle batteries. It also provides an incentive to cover the additional costs associated with the installation at the charging points of technologies designed to ensure the interaction between the vehicle and the network. Finally, the decree provides simplified modalities for the participation in UVAM for some categories of charging stations (such as domestic ones) and above all it opens the possibility to provide new types of services, such as frequency and voltage control. The decree speaks of "Virtual Units Enabled for charging" (Unità Virtuali Abilitate di Ricarica, UVAR), consisting exclusively of charging points for zero emission cars.

Also in this case the ARERA has the task to implement ad hoc dispositions to integrate this novelty to the dispatching services. With the document for consultation 201/2020/R/eel the ARERA defines the first guidelines for the revision of its electrical dispatching regulation to promote the participation of electric vehicles in the MSD, through charging infrastructures equipped with V2G technology. In the document, rules are indicated to define how to include new generations of V2G technology charging columns in advanced dispatching projects, in order to contribute to the balancing of the electrical system. The proposal to reduce the minimum power to be made available upstream and/or downstream from 1 to 0.2 MW has been accepted and applied to all UVAM in order to preserve the principle of technological neutrality. With regard to the coverage of additional costs of the charging infrastructure for the provision of ancillary services provided by the Decree, a decision has not yet been taken, but in the document for consultation ARERA anticipates that these rules will be defined as soon as the Italian Electrotechnical Committee (CEI) has identified the minimum technical requirements that must have the devices and meters installed at the connection point (even when integrated into the charging infrastructure) for the provision of ancillary services.

Considering that in Italy there are about 40,000 electric cars (estimate end 2020), assuming that 10% of these cars are connected to the grid for charging in the evening and night and assuming an average size of 10 kW for each column, the electrical system could count on 40 MW of additional capacity to be managed with increasing /decreasing services. Of particular interest are the public/private companies that own hundreds of vehicles; think, for example, of the large electric company fleets that, connected with V2G systems to the grid, could provide some MW of capacity for each one. Much will also depend on how the cars are used; it is clear that the most suitable type of charging for V2G modes is the slow charging performed in the evening or at night, using private domestic or corporate columns. Fast recharging, on the contrary, requires that priority is always given to vehicle refueling so it lends itself in a very limited way to V2G applications.

In general, the UVAM project as well as the V2G applications are proceeding quite well, constituting interesting mechanisms for achieving the objectives of the Clean Energy Package to 2030. The opening of the MDS to new users, allows the reduction of the use of conventional power plants for the regulation of power flows on the grid and the provision of ancillary services with the consequent reduction of CO_2 emissions and the incentive of users to install RES plants. In addition, the possibility of being able to use own electric vehicle to help the grid in the ancillary services in return for remuneration is a way to encourage people to purchase electric vehicles, which will not only contribute to the reduction of emissions, but will also have an economic return.

2.5.2 Renewable Energy Communities

The EU Renewable Energy Sources Directive 2018/2001 (RED-II) defines the role of prosumers and Renewable Energy Communities (REC) with the objective of growing the decentralisation of electricity generation in the scenario of the liberalisation of the energy market. RECs exist since many years in some of the northern Europe countries, such as Denmark and Germany. In these communities, renewable energy is generated by facilities owned by the participants that are producers, consumers and prosumers. The aim of the community is to provide environmental, economic and social benefits to the members rather than financial profits. Inside a REC, it is allowed to produce, store, consume and sell renewable energy generated within the community. For a REC it is possible to participate in all electricity markets, either directly or by means of aggregation. Members of a REC are users belonging to the same building or group of closeby buildings. By definition, there is indeed a requirement about closeness of the members of a Renewable Energy Community. The holders of points of delivery of electricity at low voltage can be members of a REC. Moreover, they need to stay downstream the same MV/LV substation transformer, so that their participation in the REC does not imply any change and/or upgrade of the existing electricity network. In a REC, there is commonly one of the members (a representative of the community) who sends the application to the authority to get the expected benefits from the self-consumption of the renewable energy generated inside the community. The authority, on the basis of the measured consumption/generation data collected at the points of delivery or common coupling, assesses the collectively self-consumed energy and grants incentives to the representative of the community. The latter is normally the owner of one or more renewable energy production plants within the community. As an example, in Italy, the Decree 162/19 (Milleproroghe Decree), transposing the European RED-II Directive, states that the REC representative must submit a request to the authority (GSE) with all the information that is necessary to identify the energy community [22]. The Decree supports and encourages self-consumption of energy generated from renewable energy production plants, which can also be delayed through storage systems. The representative of the Community, who also receives the economic benefit, is the owner of one of the renewable energy plants installed to constitute the community. The renewable generators must individually have a rated power less than or equal to 200 kW and must have entered into operation after 01/03/2020 and within the following sixty days from the date of entry into force of the transposition of Directive 2018/2001/EU (30 June 2021 as final date). The community downstream a MV/LV substation can include both production plants and utilization plants. The GSE then recognizes an incentive equal, on an hourly basis, to the sum of two terms:

1. The product between the unit amount refunded (totalling 0.822 c€/kWh for the year 2020) and an amount of energy equal to the minimum between the electrical energy supplied by the renewable energy installations covered by Decree 162/19 and the total electrical energy consumed at the Point Of Delivery owned by members of the same REC.
2. The product between the coefficient of avoided losses (1.2% in the case of production plants connected to medium voltage networks or 2.6% in the case of production plants connected to low voltage networks), the hourly zonal price and a quantity of energy equal to the minimum between the electrical energy injected by the plants covered by Decree 162/19 and the total electrical energy consumed at the Point Of Delivery owned by members of the same REC.

As already said, this incentive is granted to the representative of the community, in turn he then distributes it to the other members of the community who have contributed to get the incentive. The first item of the two above listed indeed is as larger as the minimum between injected and consumed electricity is higher, that is to say, as the effect of levelling (in terms of energy) is increased. As a result, the more the members consume at the time the renewable energy plants inject, the better it is. Furthermore, the self-consumption of energy lets the attainment of many benefits both for the user of the energy community and for the grid operator. The user will gain an economic return due to the energy not purchased plus the incentive, while the grid operator will have as main benefit the reduction of power flows in the power lines upstream with a resulting reduction of power losses. Moreover, the self-consumption of electricity on site could, in the longer term, reduce the need to upgrade existing networks infrastructures or to build new networks, as it would contribute to the reduction of the maximum power required at the connection points or points of delivery. A further benefit is the optimisation of the use of substations and connection points, reducing connection costs. However, in Italy the latter two benefits are not fully operational due to the fact that the incentives are referred to hourly energy (and not power) balance. Consequently, these latter advantages can only be accounted for in a future reference frame.

The blockchain technology implementation for supporting the RECs deployment would in first place allow full transparency of the incentive program for self-consumption. Users would be identified on the blockchain platform as members of the REC and the hourly incentive program described above could be supported by suitable SCs, in charge of distributing the benefits to all those that have contributed to the hourly energy balancing. In this way, energy transactions among citizens could be managed by a blockchain based platform, while ensuring privacy. The blockchain would ease the spontaneous aggregation of loads and producers without an intermediary, namely a representative of the community, as all users could stay in direct contact with the authority. The latter, hour by hour, by means of the smart contract could recognise the incentives directly to the users while acquiring the required technical objectives.

References

1. P. Faria, J. Spínola, Z. Vale, Methods for aggregation and remuneration of distributed energy resources. Appl. Sci. **8**, 1283 (2018). https://doi.org/10.3390/app8081283
2. Z. Ma, J.D. Billanes, B.N. Jørgensen, Aggregation potentials for buildings-business models of demand response and virtual power plants. Energies **10**(10), 1–19 (2017). [Online]. Available https://ideas.repec.org/a/gam/jeners/v10y2017i10p1646-d115763.html
3. European Union Emission Trading Scheme, *Demand Side Response Aggregator*. Accessed 9 July 2020. [Online]. Available https://www.emissions-euets.com/internal-electricity-market-glossary/855-demand-side-response-aggregator-dsr-aggregator
4. F. Grosso, *Esperienza nel ruolo di aggregatore: focus progetti pilota in Italia*. Accessed 18 Aug 2020. [Online]. Available http://aeit.diees.unict.it/down2018/workshop_23nov18/07_EnelX-Catania23_11_18.pdf
5. L. Marchisio, F. Genovese, F. Ruffo, *L'apertura delle risorse distribuite al mercato dei servizi: quale bilancio?* Accessed 24 Aug 2020. [Online]. Available https://download.terna.it/terna/0000/1224/89.PDF
6. G. dei Mercati Energetici, *ME—info societarie* (2020). Accessed 1 Oct 2020. [Online]. Available https://www.mercatoelettrico.org/it/GME/Info/ProfiloAziendale.aspx
7. The European Parliament and the Council of the European Union, Regulation (eu) 2019/943 of the European parliament and of the council on the internal market for electricity, ed. by E. Parliament (2019). Accessed 25 Aug 2020. [Online]. Available https://eur-lex.europa.eu/legal-content/EN/TXT/?uri=CELEX
8. A. Galliani (ARERA), *Evoluzione del mix produttivo di energia elettrica e suoi effetti sul sistema*. Accessed 24 Aug 2020. [Online]. Available http://www.tpg.unige.it/TPG/wp-content/uploads/2015/12/Galliani_AEEGSI_Genova_30ottobre2017.pdf
9. ARERA, *Verso la fine del mercato tutelato*. Accessed 31 Aug 2020. [Online]. Available https://www.arera.it/it/consumatori/finetutela.htm#
10. N. Carrieri, *Filiera dell'energia elettrica, dalla produzione al contatore*. Accessed 31 Aug 2020. [Online]. Available https://luce-gas.it/guida/mercato/filiera-elettrica
11. Enerchain, *Enerchain—Decentrally Traded Decentral Energy*. Accessed 23 Oct 2020. [Online]. Available https://enerchain.ponton.de/

12. C. Zhang, J. Wu, C. Long, M. Cheng, Review of existing peer-topeer energy trading projects. Energy Procedia **105**, 2563–2568 (2017), in *8th International Conference on Applied Energy, ICAE2016*, 8–11 Oct 2016, Beijing, China. ISSN: 1876-6102. https://doi.org/10.1016/j.egypro.2017.03.737. [Online]. Available http://www.sciencedirect.com/science/article/pii/S1876610217308007
13. EM Compass, *Using Blockchain to Enable Cleaner, Modern Energy Systems in Emerging Markets*. Accessed 23 Oct 2020. [Online]. Available https://www.ifc.org/wps/wcm/connect/46ad7055-a5b5-4db0-af78-92fc67a61566/EMCompass-Note-61-Blockchain.pdf?MOD=AJPERES&CVID=mthzuiy
14. M. Andoni, V. Robu, D. Flynn, S. Abram, D. Geach, D. Jenkins, P. McCallum, A. Peacock, Blockchain technology in the energy sector: a systematic review of challenges and opportunities. Renew. Sustain. Energy Rev. **100**, 143–174 (2019). ISSN: 1364-0321. https://doi.org/10.1016/j.rser.2018.10.014. [Online]. Available http://www.sciencedirect.com/science/article/pii/S1364032118307184
15. D. Loghin, G. Chen, T. Dinh, B. Ooi, Y. Teo, *Blockchain Goes Green? An Analysis of Blockchain on Low-Power Nodes*. Accessed 28 Oct 2020. [Online]. Available https://www.researchgate.net/publication/333160699_Blockchain_Goes_Green_An_Analysis_of_Blockchain_on_Low-Power_Nodes
16. D. Carboni, *Incognite e attrattive della blockchain nel settore energetico*. Accessed 27 Oct 2020. [Online]. Available https://www.agi.it/blog-italia/digitale/incognite_e_attrattive_della_blockchain-3490185/post/2018-02-14/
17. M. Buchmann, *Blockchain Technology—A Threat to Distribution Network Operators?* Accessed 27 Oct 2020. [Online]. Available https://energypost.eu/blockchain-technology-a-threat-to-distribution-network-operators/
18. ARERA, *Delibera 05 maggio 2017 300/2017/r/eel*. Accessed 28 Oct 2020. [Online]. Available https://www.arera.it/it/docs/17/300-17.htm
19. Terna, *Regolamento recante le modalità per la creazione, qualificazione e gestione di unità virtuali abilitate miste (UVAM) al mercato dei servizi di dispacciamento—regolamento msd uvam*. Accessed 28 Oct 2020. [Online]. Available https://download.terna.it/terna/Regolamento_UVAM_MSD_8d803bd91884d36.pdf
20. Terna Spa, *Five Things You Need to Know About UVAM*. Accessed 28 Oct 2020. [Online]. Available https://lightbox.terna.it/en/distributed-resources-UVAM
21. Ministero dello Sviluppo Economico, *Decreto 30 gennaio 2020: Criteri e modalita' per favorire la diffusione della tecnologia di integrazione tra I veicoli elettrici e la rete elettrica, denominata vehicle to grid*. Accessed 29 Oct 2020. [Online]. Available https://www.gazzettaufficiale.it/eli/id/2020/02/14/20A00891/sg
22. ARERA, *Documento per la consultazione 112/2020/r/eel, orientamento per la regolazione delle partite economicche relative all'energia elettrica oggetto di autoconsumo collettico o di condivisione nell'ambito di comunità di energia rinnovabile*. Accessed 28 Oct 2020. [Online]. Available https://www.arera.it/allegati/docs/20/112-20.pdf

Chapter 3
Blockchain Experiences in Power Systems

G. Sciumè, E. Riva Sanseverino, and P. Gallo

Abstract The blockchain technology is recently gaining a huge popularity due to a wide variety of use cases, where it can be adopted, and the benefits it provides. As already said, the widespread of renewable and distributed generation requires a rethinking of the management of the energy flows in the power system. The blockchain technology in the energy field is now a realistic perspective, particularly for microgrids, in which energy production is intrinsically distributed. The decentralized structure of a blockchain can be perfectly adapted to the control of production/consumption/storage within microgrids. It is thus understood that this technology can guarantee the traceability and security of energy transactions. As introduced in the first chapter, in recent times in this sector the blockchain applications are mainly oriented at P2P electricity trading, electric vehicles or DR. In this chapter, the features of blockchain technology that are useful for the energy system are addressed in more detail looking at the main applications in the energy field and the suitability of the blockchain for these applications.

3.1 Features of the Blockchain for Power Systems Applications

It can be easily understood that there is a difference between blockchain applications that only operate on digital assets and blockchain applications that require the interfacing with physical objects. This is also the case of the electrical energy application field, for which physical electricity transactions must be managed through smart meters and actuators. Else than a cryptographic signature or a digital recognition of a virtual item, in this case, there must be a "certification" for a physical measure or control system, as well as for the software controlling the actuators or other operations in the virtual dimension. As a matter of fact, in the physical world, also complex applications (such as SCs) are needed. These may provide a set of execution commands allowing to adhere to predetermined workflows. In the case of

G. Sciumè (✉) · E. Riva Sanseverino · P. Gallo
University of Palermo, Viale delle Scienze, Building 9, 90128 Palermo, Italy
e-mail: giuseppe.sciume01@unipa.it

G. Sciumè et al. (eds.), *A Practical Guide to Trading and Tracing for the Energy Blockchain*, UNIPA Springer Series, https://doi.org/10.1007/978-3-030-96607-2_3

P2P trading, the triggering event is connected both to the availability of money, from one side, and to the availability of energy in the considered time frame, from the other. In the electrical energy field, these conditions are needed but not sufficient for any transaction to take place. There are some technical requirements that are connected to the capacity of the connections and relevant regulatory issues that should be considered. Such conditions cannot be verified in a distributed way in many cases and this calls for the action of DSOs, whose role is fundamental to ensure efficient integration of this technology into the network. For this reason, the permissioned blockchains may be more adequate to physical world applications. This is even more true in the power systems field where physical objects belonging to a third party are involved in the transactions. In these blockchains, only a few "permissioned" nodes can generate and write blocks. These nodes are not anonymous and they may implement different functions. With respect to this issue, it is advisable that, as an example, one of the blockchain nodes is owned by DSO. The latter can supervise transactions and provides a certification about their technical feasibility [1]. A "validation" procedure that can take place between steps 2 and 3 in Fig. 1.2 supports this certification. As an example, if the validation of electricity transactions requires to run an Optimal Power Flow or similar algorithm for which network's electrical data must be known, for the time being, only the DSO can run it. Moreover, in the energy field, the timing and cost of transactions is a significant problem. The use of permissionless blockchains instead provides greater security if data can be released publicly. However, due to privacy and the computational time of the heavier consensus algorithms, permissionless blockchains appears not to be adequate for handling energy transactions. For this reason, energy-related applications are more efficiently handled using permissioned blockchains, despite in the literature there are a number of solutions that use permissionless ones. Another challenge is about using the blockchain for aggregation purposes and for balancing. The NEMoGrid European project [2] involving various companies and research organizations from Germany, Switzerland and Sweden, analyzes the economic and technical impact of electricity trading between households within an area, i.e. when selling solar power generated locally to other households. In this way, an insight into the mechanism of price variation in the presence of a P2P business model is gained as well as the suitability of blockchain technology to support flexibility and grid stability issues through aggregation in small pilot regions. In the electricity field, a number of applications deal with the emission of utility tokens. The latter are digital tokens of cryptocurrency that are issued in order to fund the development of the cryptocurrency and that can be lately used to acquire a good or service put forward by the issuer of the same cryptocurrency. As a consequence, these cryptocurrencies prove the ownership of an asset/privilege. As an example, tokens can be exchanged as a counterparty of (green) electricity generation. So, in this case, if one party generates green electricity, it is then remunerated by a SC and is shared on the blockchain with a utility token. The latter can then be exchanged with security tokens also called securities, like bitcoin, storing a value. Another important feature is the flexibility in supporting SCs or other forms of logic. Indeed, the possibility to implement SCs and not only "local" client applications that do not change the state of the blockchain is a very relevant feature of decentralized computational resources in blockchain platforms. SCs can process digital transactions and can be used to guarantee that a specific workflow is met. In all the applications concerning power systems, there is a need for SCs. In all cases, where a technical verification over the infrastructure is needed, the SC needs to run power analysis or power optimization codes. As an alternative, the SC can check some metadata sent as inputs of the SC. The difference is that in the former case all nodes run the SC to verify the validity of the transaction, while in the latter case there is a trusted authority that must issue a "permission", used under the form of metadata. The mechanism of SC execution, delivering e-tokens if given flexibility at node A is provided, is depicted in Fig. 3.1. The output of the SC is the feasibility of the transaction and thus the consensus about starting the actuation of devices and the transfer of digital assets.

Moreover, the decentralization of the network, the transparency of exchanged data and the possibility to implement transactions without the intervention of third parties allows to avoid market abuse, reduce transaction costs and create innovative business models for the electricity market directly involving end users.

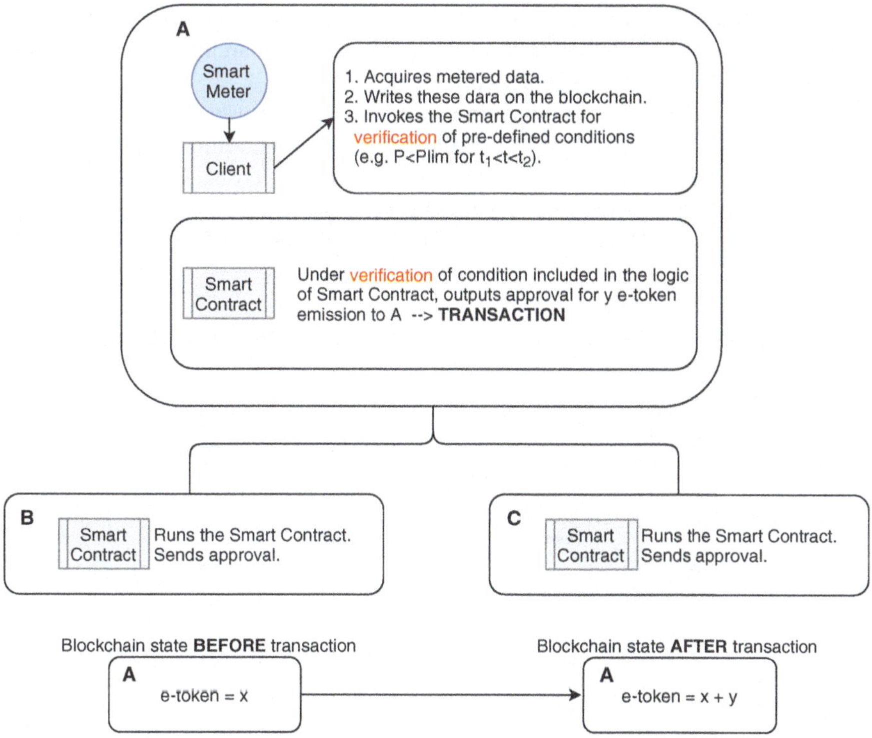

Fig. 3.1 Example of a smart contract execution

3.2 Peer-to-Peer Energy Exchanges

3.2.1 Peer-to-Peer Principles

A Peer-to-Peer system, in general, is a system consisting of a computer network in which the computers of connected users act at the same time as clients and servers. It is a distributed network that works with the principles of a centralized client-server network. In this way, users are able to directly access each other's computers, viewing and retrieving files in mass storage and making the files they wish to share available. Each of the computers of the users who are part of the P2P network downloads, keeps the exchanged files stored and, above all, shares them with other users. In these systems the simultaneous use of all the physical locations in the computers where the same file is located is managed, optimizing the use of bandwidth in proportion to the number of people connected at a given time, so as to reduce the time of downloading files. Thus, a user is able to download portions of the same file from multiple users in proportion to the bandwidth available to each user who owns the file. In relation to the file to download, is named *seed* the user who owns a full copy of the file object of transaction and *leach* the user who has not yet finished downloading a full copy of the same. Figure 3.2 shows the comparison between the architecture of a classic client-server network and a P2P network.

There is no standard communication speed, since each user receives and sends what the bandwidth of his ADSL or fiber line allows him. Among the advantages it can be immediately noticed the absence of hardware costs, such as dedicated servers. This also means that file download speeds can be much faster, in relation to the number of files searched by users connected to the network. A problem to take

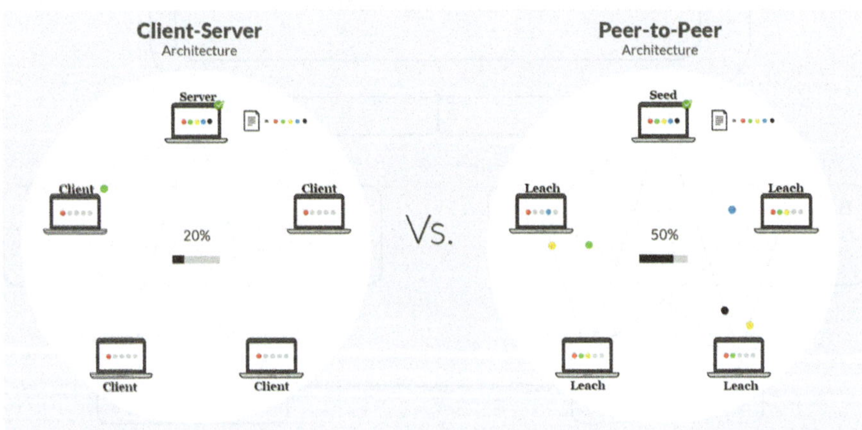

Fig. 3.2 Client-server architecture versus P2P architecture

into account is the security of the network user, as there are serious risks of external attacks, if the router is not adequately protected (for example through a firewall).

The blockchain, which is a distributed network like P2P networks, allows to overcome this problem through the use of encryption and consensus algorithms. In addition, as already mentioned in the introduction chapter, each node of the blockchain contains all the info exchanged on the network and how for P2P networks works both client and server. To date P2P networks have been used to exchange video and audio files, such as Torrent, Gnutella and Napster. But thanks to blockchain the principle of P2P networks can be applied to different realities, including energy exchanges.

To date, in Italy, a prosumer can only inject excess energy produced into the grid and access a service called "on-site exchange". The on-site exchange service is a particular mode of on-site self-consumption that allows the electricity produced and then injected into the grid at a certain time to be compensated for the electricity absorbed and consumed at a later time. In the on-site exchange, the power system is used as a means for the virtual storage of electricity produced but not simultaneously self-consumed. Periodically, the GSE, through the meter readings that are communicated by the DSO, verifies how much energy has been consumed and how much injected. The GSE does this verification annually, but every quarter it makes an estimate of the advances and pays them to the prosumer. At the end of the year the estimates are corrected and the report for the past year is paid. If at the end of the year it turns out that more kW have been injected into the network than those that have been consumed, the prosumer receives a contribution which is the sum of two terms:

1. the first is a refund, for each of the kW consumed;
2. the second is a sale consideration for the kW injected and not consumed.

The value of the refund, is calculated considering a price that is not exactly the one paid by the user in the bill, but it corresponds to a 30% lower price. In fact, the taxes paid on the invoice are not refunded. It is, however, a considerable amount that is equivalent to 70% of the bills paid. The value of the second item, that is the consideration for the kW injected in the grid, is valued by multiplying the injected energy by the price per kW valid for the territorial area in which the plant is active and detected in the market, in particular in the MGP. The on-site exchange model will however be outdated by the new and wider definition of Renewable Energy Communities in Italy. As it is possible to understand from the on-site exchange program, the prosumer cannot decide to whom and the amount of energy to sell, but it is all managed by the GSE and the DSO that communicates the consumption and injection data through a centralised system. The possibility to develop different business models also depends on the technology that supports them. In this perspective, blockchain technology thanks to its features, enables the possibility of implementing an energy transaction directly between "peers", in the form of an exchange or sale of the energy surplus to other user of the power network, thus P2P energy exchanges (see Fig. 3.3).

The blockchain can be used as an immutable distributed ledger of energy transactions capable of certifying the energy exchanged between two users without the need for a trusted third party, such as the GSE or DSO, thus providing the conditions

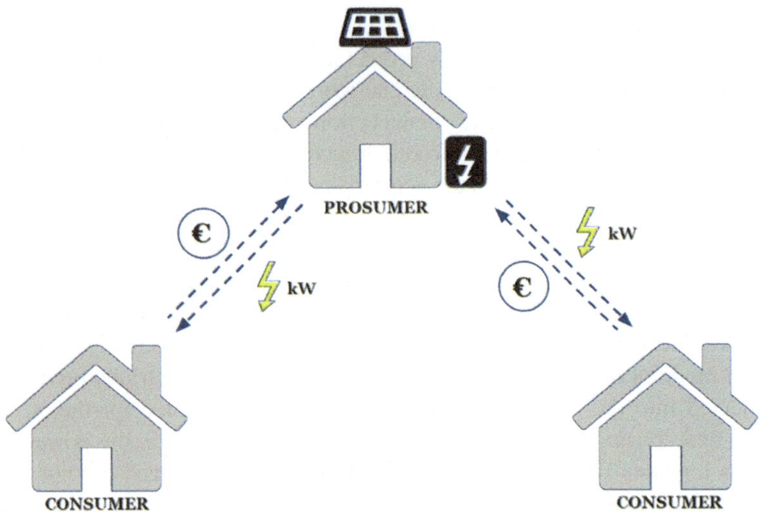

Fig. 3.3 P2P energy exchange

to create a P2P energy market. With this system, the prosumers can either sell their surplus energy in the market or stipulate and manage contracts with other end users.

3.2.2 Actors of the P2P

A P2P energy transaction directly involves two users, a producer/prosumer selling and a consumer/prosumer buying. Depending on the scenario, the buyer can be a company that buys energy in the wholesale market and that sells it back in the retail market or simply an end user that consumes the energy purchased from the neighbor. However, it must be kept in mind that an energy transaction indirectly involves the system operator. In the low voltage grid, the system operator has no decision-making possibility over the execution of the transaction, as LV networks are generally designed and implemented considering a unitary contemporaneity coefficient. However the system operator must ensure that the transaction is physically possible, so that the grid is efficient. This can be guaranteed by paying a fee to the system operator for each transaction executed, exactly as it happens for the system charges of the energy bill. For transactions occurring in the High or Medium voltage grids, the system operator manages a more delicate balancing situation and, based on load flows analysis, it can give (or not give) the permission for the execution of transactions. This analysis can be performed on-chain through for example a SC execution or off-chain through dedicated software (more realistic). Finally, in blockchain platforms there is the blockchain network manager, who is also responsible for users' identities. The latter may be either a specific company or the same energy retailer that already has dealings with the user.

3.2.3 P2P Benefits

P2P energy exchanges give rise to an energy ecosystem that has the potential not only to provide economic and environmental benefits, but also to improve grid reliability and production plant efficiency. The fragmentation of the energy market into local markets through the introduction of many decentralised energy sources also increases competition bringing even further benefit of consumers. Increased competition has already started the energy transition by shifting the energy production market from fossil fuels to renewable energies. This market transformation would not only bring an ecological benefit, but also has a real financial advantage because renewable energy production is getting more and more cheap. The creation of a P2P energy market can also contribute to the balancing of the power grid by reducing, for example, load peaks. By aggregating different users, prosumers and consumers, it is possible to create energy communities that, acting in a coordinated way, can contribute to the balancing of the network. More in general for end users the benefits are economic as they can choose who to sell their surplus energy or from whom and when to buy the energy. This flexibility to sell and buy could also encourage end users to install small renewable energy plants. Regarding the system operator, the benefits are especially in terms of grid management. Thanks to local markets, during peak hours small communities can self-supply and release capacity of the upstream grid with the consequence of reducing the need to expand power lines and the installation of new production plants to meet the energy demand. In addition, using blockchain for P2P energy exchanges it is possible to achieve benefits in terms of:

- *Optimization*: better monitoring of consumption, quantity and price control.
- *Transaction costs*: potential cost reduction for the presence of third parties (e.g. brokers).
- *Exchange between individuals*: the user will be able to sell his energy.
- *Automatic execution*: use of SCs to reduce time, costs and control activities.
- *Security*: information distributed over several nodes, difficulty of computer attack.
- *Wider choice*: the user can choose to consume clean energy by choosing the system from which he wants to buy.

3.2.4 State of Art

In the literature there are many works focusing on this application and proposing a blockchain-based framework to support and promote P2P electricity exchanges among end users (producers and consumers) of a microgrid [3–7]. As an example, the reserach in [6] proposes an blockchain that employs hyperledger fabric to implement a P2P market, so as to handle the increasing volume of renewable energy with Energy Storage System (ESS) installed at Low Voltage. The work in [7] suggests the use of an energy trading digital platform for ESS management and energy tokens exchange to solve the security problem through a blockchain private network. It is thus implemented a P2P market by executing automatic transactions without involving third parties supported by SCs in a transparent and immutable way. The proposed architecture consists of a private blockchain network in which energy is exchanged through ESS and energy tokens through the blockchain. Users register sales and purchase offers on the blockchain, after which the Distribution System Operator (DSO) compares the total amount of the purchase and sale. Trade is realized if the amount of energy to buy and sell is equal. So the SC moves the token and the energy according to the accomplished trade, and then the result is recorded on the blockchain.

Several companies have instead developed applications in this sector. The first P2P blockchain application is the so-called "Brooklyn Microgrid" developed in New York, USA, supported by Siemens Digital Grid and LO3 Energy [8, 9]. The first installations date back to 2016 and the platform allows the prosumers to sell the excess electricity directly using SCs based on Ethereum and pBFT consensus, implemented on Tendermint. The first process included 5 prosumers and 5 nearby consumers and brought to the first energy transaction ever registered on a blockchain worldwide. The energy surplus was registered by smart meters suitably designed for managing energy data and measurements. The latter are thus translated into energy tokens that can be exchanged in the local market. The digital token represents a given amount of energy that has been produced by the photovoltaic plant, in this way, it can be transferred from the portfolio of a consumer to the end-users by the blockchain technology. The microgrid users interact with the platform specifying their individual price preferences under the form of availability to pay or sell electricity. The platform can visualize energy prices in real-time. The public register saves the clauses of the digital contract, the identity if the parties that implement the agreement, the volumes of energy injected and the energy consumption measured by the smart devices as well as the chronological arrangement of the transactions. Any member of the community can have access to all historical transactions in the log and verify the connections. In the initial phase of the project, the users could manually act on the platform, they could for example make an offer for a given price and a given hour. Now, in the second phase of development, the platform manages automatically a lot of functions and hosts more than 300 domestic users and small business users, including about 50 prosumers equipped with photovoltaic installations and a small wind producer. The Brooklyn MicroGrid project serves as a testbed for the development of new business models that promote consumer involvement in energy community projects. Local energy exchanges open up the possibility of saving on energy costs. After the Brooklyn MicroGrid project, other companies around the world have started developing P2P energy trading projects. In 2017, in Hamburg was founded the company Enyway [10] with the mission of facilitating a democratic use of energy through the connection between producers and consumers. The business model allows consumers to buy electricity from individual owners of a wind turbine or a solar roof. Users can also choose the projects from which they can buy energy in Germany, with a nice interface made of images and data, such as in the case of apartments/rooms offered on the AirBnb platform. In France, the Sunchain [11] experience is expanding. It allows energy exchanges between prosumers and consumers using a blockchain platform not associated with a crypto-currency. Power Ledger [12] is an Australian company that has developed a renewable energy trading platform enabled by blockchain technology. Since November 2018, the Australian Power Ledger blockchain has been employed to track solar energy transactions between a small set of 18 families in the city of Fremantle, Western Australia. Residents could set their own prices and sell solar energy generated from their roofs to the neighbors. The same Power Ledger blockchain in Japan is being used for Renewable Energy Certificates (for solar power generation) issueing and is recently enarging to the whole SouthEast Asia, supported by a partnership with the renewable energy business BCPG and Thai Digital Energy Development (TDED). The so-named Non-Fossil Value credits let electricity retailers to get certificates of renewable energy for electricity that has been fed back into the grid by Photovoltaic systems. Power Ledger's decentralized accounting technology ensures that every transaction is immutably registered in real-time, in order to discourage renewable energy certificates from being used more than once.

3.3 Vehicle-to-Grid

3.3.1 Vehicle-to-Grid Principles

Electric vehicles are a natural field of application for blockchain technology. The new mobility concept, aiming increasingly towards a "digital" model, inevitably requires improved safety and reliability standards. According to [13], financial services, supply chain and mobile are three of blockchain's main areas of application. In addition to facilitating transactions between vehicles, SCs and real-time data exchange, the implementation of blockchain-based solutions could solve today's difficult problems such as the traceability of original spare parts to identify counterfeit parts or traceability of vehicle consumption to facilitate recovery at the end of its life. In this sense, the blockchain offers a more secure and transparent way to communicate thanks to the continuous sharing of information between nodes of the same and the increased difficulty of tampering with individual blocks. The expected arrival of millions of battery-powered electric vehicles in the coming years represents a major challenge for the power grid. At the same time, however, the electric car also provides very interesting opportunities to improve the management of electrical systems without having to invest unsustainable amounts of money to adapt the distribution infrastructure [14]. In this view the Vehicle-to-X (V2X) concept was conceived, which refers to an exchange between a vehicle and another element of a power grid. Electric vehicles can be considered as mobile accumulators capable of exchanging energy with the grid, other electric vehicles or the own home. This leads to four scenar-

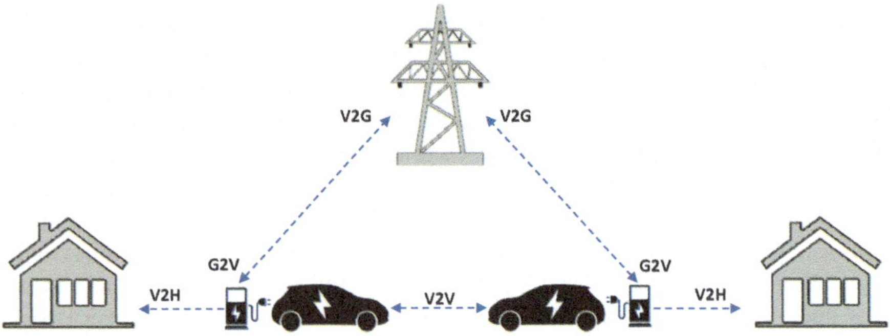

Fig. 3.4 V2X scenarios

ios: Vehicle-to-Grid (V2G), Grid-to-Vehicle (G2V), Vehicle-to-Vehicle (V2V) and Vehicle-to-Home (V2H). Figure 3.4 shows a general representation of these four scenarios.

V2G requires cars to act as accumulators so that they can be converted into reserves at critical times to stabilize the grid and avoid overloading. Underlying this mechanism is an energy exchange process managed by a power inverter operating in both directions (low voltage for the grid and high voltage for the battery), whose operation is optimized according to the needs of each component of the reference grid. The bi-directional inverter is able to take energy from the grid to charge the battery (G2V), or supply energy to the grid by taking it from the battery (V2G), see Fig. 3.4. All this is managed by a control unit that must meet the requirements of the network operator, respecting the minimum charge level and full charge time will provided by the owner of the vehicle. Depending on the commands it receives from a network operator, the bi-directional inverter either takes energy from the grid to charge the battery or takes energy from the battery to send it to the grid like any generator. The management software is designed to provide the intended service (according to the command of the network operator for V2G, according to the maximum utility for the user for V2H).

Considering that an electric car spends several hours a day parked in a public parking or in a private garage, in many cases connected to a charging column, it is evident how this power can be used, for part of the time, as a resource for the electrical system to balance the need for energy at all times with the production from renewable sources not programmable. The V2G provides for the exchange with the network of energy with the aim of increasing or reducing the overall power absorbed by the electrical system. The V2H provides the same thing, but instead of the electricity network there is the home and the goal is to have maximum savings in the bill for the end customer. In this case it is the owner of the house or the energy management system that intends to increase self-consumption, to make the most of self-production or reduce the peaks of power taken, saving on costs related to the power involved.

V2V applications, on the other hand, concern the possibility of transferring energy from one vehicle to another through the use of a cable or even wireless. Due to the distribution network not yet fully equipped with charging columns scattered all over the territory, in case of battery discharge far from the charging points, until today, the only solution was to contact a tow truck and be transported to the nearest charging column. The V2V system allows electric cars to communicate with each other and enable the exchange of energy between two vehicles. However, it is necessary to upgrade the on-board power electronics so as to allow two-way flow of energy, which currently only goes from the energy source to the engine.

Distributed technologies and Blockchain therefore become ideal for managing systems like those described that, having to rely on the exchange of resources and information in an environment without hierarchies, is by its nature P2P. The use of blockchain in V2X can be a guarantee of efficiency and reliability of communication combined with transparency and security of information. In addition, the use of SCs would allow to implement the logic for network support services provided by electric vehicles. In this case, charging the car represents a service that the user offers to the network and for which he will be remunerated with energy tokens. In this scenario, electric vehicles can contribute to DR, improve grid resilience and reduce peak loads through charging and discharging. The Blockchain allows to verify and record all transactions that take place in a V2X network, ensuring its correct operation and allowing, at the same time, to acquire data useful for its continuous improvement. Applications based on the Blockchain allow people to make their charging stations available to the public in moments of inactivity, receiving tokens in exchange, so drivers of electric vehicles can search for the nearest charging stations by refueling their batteries before they discharge.

3.3.2 Actors of the V2G

As already said, the V2G is a technology that enables the exchange of energy between the electric vehicle and the network, own home or another electric vehicle. The potential of this technology is really high because with an adequate implementation of the infrastructure, it will be possible to have millions of electric vehicles that will act as mobile energy assets and that will not only support local networks, but also contribute to the national transmission system. Typically, commercially available electric vehicles are equipped with battery packs ranging between 30 and 50 kWh, with true autonomy ranging between 200 and 400 km. That said, it is clear that the contribution that a single car can provide to the power grid is not very high, since it is necessary that each vehicle does not fall below a certain charge limit, necessary both for the health of the battery and for the owners' needs. It is necessary a population of electric vehicle batteries that provides, in a coordinated way, both the power and the energy necessary to provide a significant contribution in an area of the electricity market. For this reason, in order for electric vehicles to contribute to the needs of the power grid it is necessary that they are aggregated and properly managed. This role is fulfilled, as for the DR, by the aggregator, which aims to aggregate vehicles that do not belong to them by managing their loading/unloading cycles to sell services on the energy market, representing the point of contact between the owners of electric cars and the system operator. The authors in [15] identify 12 possible actors for V2G applications, but in general the main actors are the system operator, the aggregator and the electric vehicle. The electric vehicle is identified with the owner of the vehicle itself, the system operator can be the TSO or DSO, while the aggregator can be different depending on the business model. For more details about business models for V2G please refer to [15]. According to [16] the main players, even the most sensible, who can take on the role of aggregator for V2G applications can be:

- Delivery company, car sharig or any company that deals with transport and offers mobility services to customers. This type of actors are called "fleet managers" and have the role of managing the availability of the fleet and then sell its services to the market or to an operator. In this case the fleet is connected to a single network point, just think of the car sharig parking lots or the garage of a delivery company. As fleet owner can decide to load electric vehicles during off-peak hours and sell the energy stored by the vehicles during peak hours contributing to the reduction of network overload.
- The energy retailer. In this case the aggregator is the owner of a fleet distributed in the territory who can choose to manage the vehicles in order to provide services to the system operator. In this case the aggregator must have as many electric vehicles in its program as possible in order to generate revenue from the sale of network services.
- Battery manufacturer. In this case the owner of the electric vehicle chooses to transfer, in exchange for a reduced rental fee, the ownership of the battery to the battery manufacturer, who manages the battery and has full responsibility for possible damage caused to the battery as a result of providing V2G services.
- An operator specifically created to manage the charging points and communicate with the owners of electric vehicles to control the charges and discharges. This figure is called Electric Recharge Grid Operator (ERGO).
- Communication company. The aggregator is a platform provider that aggregates electric vehicles to enable them to participate in energy markets and services such as frequency response. It acts as a bridge between needs and available resources (electric vehicles, batteries, charging stations).

In blockchain applications, the aggregator, regardless of its identity, could be the platform manager. In this case, the blockchain would contribute to the exchange of data in real time between the vehicles and the system operator and to the remuneration of users following the service provided in a transparent and automatic way, for example by using SCs.

3.3.3 V2G Benefits

V2X applications allow to have a "widespread battery" available that can balance supply and demand of energy in cases of sudden lack of generation (due to a cloud passing on a Photovoltaic field) or of load (due to an outage in the grid). The benefits of V2G and G2V applications are thus deployed in the electrical system. In both cases, the electric vehicle is considered as a mobile battery that can interact with the electricity grid allowing the stabilisation of power flows. For this purpose, the vehicle, by storing energy at the lowest peak consumption times and returning it when needed, could provide ancillary services of frequency regulation (primary, secondary, tertiary) and real-time balancing for the electrical system when needed [17, 18]. These services are different based on the time-scales in which they are explicated (the primary starts within 30 s from the time in which the causal event takes place, the secondary and tertiary respectively within minutes and hours) and based on the way in which they are activated (the primary changes the power input as a result of the frequency measured locally, the other services are controlled by signals sent centrally by the TSO). The availability of this resource, today inevitably limited, but in the perspective of great capacity, creates a new offer of services that, in the market mechanism that regulates ancillary services, tends to reduce prices, and therefore makes less expensive the electricity for customers. Moreover, these applications will certainly decrease the use of fossil fuels in thermal electric power plants, which, otherwise, would end up in many cases to stay in service for the only purpose of ensuring the reserve margins necessary for a stable operation of the power network. V2H applications, on the other hand, could favour end users who own a photovoltaic system and who want to be self-sufficient and make maximum use of the energy they produce, reducing to minimum the cost of buying from the grid [18]. Electric vehicles could replace expensive stationary storage systems to accumulate the excess energy produced by renewable sources and shifting it to hours when consumption is high but there is no or limited sunlight. The installation of a dedicated storage system for the photovoltaic system involves an initial expense that can be an unprofitable investment that may discourage the customer. The user who has chosen to purchase an electric vehicle, instead, will find, at no additional cost, a storage of available energy, at least for a certain number of hours per day. The benefits thus obtained result in a lower net cost of managing the vehicle. It is possible to think that the user in this situation could also make a deal with a commercial entity (the aggregator) and combine the V2H operation with a V2G service to the system. V2H technology can be useful even in the absence of renewable generation: the battery of the electric vehicle can be recharged at home during the night, when energy costs less and in the morning, when it costs more, some of this stored energy can be used for domestic use. Again, at night the electric car will recharge to start the cycle again the following morning. Regarding V2V applications, the benefits are mostly in terms of management, as users would not need to reach a charging station but could recharge from other vehicles. These scenarios can be used in support of the power grid by charging discharged vehicles from other vehicles that are in maximum charge conditions and recharging them during off-load hours. Obviously, everything must be coordinated with the needs of vehicle owners in order to always have a residual charge to meet their needs. However, V2X applications raise serious doubts about the effects these could have on battery life. Currently, in fact, their duration is estimated on the basis of a certain number of charge and discharge cycles to which corresponds a certain distance. A more frequent exchange of energy when the car is stationary could in theory accelerate its wear and tear and therefore its life, bringing it ahead of time beyond the efficiency limit of 80%, which is considered the limit below which it should be replaced.

3.3.4 State of Art

A wide literature is recently available about the integration of electric vehicles into the grid and optimally coordination of charge and discharge using blockchain technology. The authors of [19] suggest the use of a private blockchain to implement a secure pricing system for charging and discharging electric vehicles. As it can be argued, all processes are executed by means of optimal SCs to supply the electric vehicles' recharge, meet individual energy consumption preferences and maximize the utility function of the network operator. By means of the execution of specific SCs between an aggregator and the owners of electric vehicles, the transaction for the exchange of energy and cryptocurrency is performed automatically and in a secure mode. One of the most recent works on the topic [20], presents an electricity transaction model based on a blockchain with the purpose of implementing P2P transactions between electric vehicles in V2G programs. The platform used for this work is Ethereum. The V2G projects running around the world are not so many. In Italy, the firs experimentation is dated May 24th, 2019. Day in which the testing phase of a project developed by Enel X, Nissan and RSE. A dedicated control platform was monitoring and actuating the charging infrastructure installed in the test microgrid in the RSE facility thus allowing the use of Nissan LEAF cars to balance the network, modulating the energy flows [21]. From the collaboration between FCA, Terna and Engie EPS, bidirectional charging systems are being experimented at the historic Fiat plant in Mirafiori (Turin), where the largest charging system in the world is on the way to be completed. To date, 64 charging points over 32 columns with rated power of 50 kW have already been built and they will be tested using the new 500e. By the end of 2021, there will be 700 charging points available, with 25 MW of total capacity, five of which will be supplied through photovoltaic panels [22].

In Europe, the largest V2G experimental project is under development in London, in the Northumberland Park bus terminal. It is called Bus2Grid and involves about 100 battery-powered public transport [23]. This initiative will initially test the potential of 28 buses connected to the grid while they are not in use, and will be carried out by SSE Utilities Solutions Ltd, a company working in the field of renewable energy solutions, the City of London and the city's public transport company. It is estimated that in this way they will be able to cope with the peaks of consumption by feeding 1MW of energy into the network. Other projects are under development in Denmark and Germany. In 2016, the companies Nissan, Enel and Nuvve have launched in Denmark the first fully commercial V2G Hub in the world. In Copenhagen, the energy company Frederiksberg Forsyning installed 10 Enel V2G units and added 10 Nissan e-NV200, 100% electric vans to its fleet [24]. When not in use, commercial vehicles can be connected to the V2G chargers to receive or inject electricity to the national grid. Besides, through the Nuvve aggregation platform, these vehicles are transformed into effective mobile energy storage solutions, helping to stabilise the Danish electricity grid. In Germany, in 2018, the Nissan LEAF became the first 100% electric car to have passed the pre-qualification phase as a power storage system according to German TSO guidelines. By implementing all the regulatory prescriptions from the German TSO for primary power regulation, the Nissan LEAF was accredited for the first time ever as a power reserve for the power grid. Nissan, together with TenneT and The Mobility House, carried out the SINTEG pilot project related to the V2G technology to increase the share of renewable energy in Germany and to reduce CO_2 with electric vehicles [25]. As part of this project, batteries in Nissan LEAF electric cars are used to store locally produced renewable electricity in order to stabilize the grid during peak demand. More specifically, wind energy produced in northern Germany is used to charge electric vehicles in the region. At the same time, the electricity from the batteries of these vehicles is injected into the grid to prevent an increase in fossil fuel generation.

3.4 Demand Response

3.4.1 Demand Response Principles

Mismatches between production and consumption are more likely to occur because unpredictable renewable energy generation is increasing in power systems. This matter implies many and diverse technical and market consequences, like congestion that may appear in distribution systems, spikes in energy prices and finally the cutback of renewable energy plants. For instance, Fig. 3.5 presents an unbalance in the grid that can occur at certain times of the day when the generation does not meet the demand.

One of the cheapest solutions to comply with unbalances is implementing the so called DR programs. In short, "DR" consists of any program that, as an effect of a specific request from the network operator, supports end users to take actions for shifting or modifying the energy consumption profile. The actuation can be implemented either directly by operators, the so-called Balancing Responsible Parties, or by end-users who settle to act either independently or based on information collected almost in real-time [26]. DR programs are created to support end users' involvement, and their response to prices is basic to attain efficient and competitive outcomes both in the market and in the technical operation. Such as it happens in any other market, also in wholesale electricity markets the more assiduously the demand participates, the more competitive and efficient the market turns to be [27]. As already explained, due to DR programs, end-users can be remunerated for agreeing to change their consumption in a specific time-frame. Putting together a large set of electricity users can sum up to significant control over demand, giving grid operators new modes to

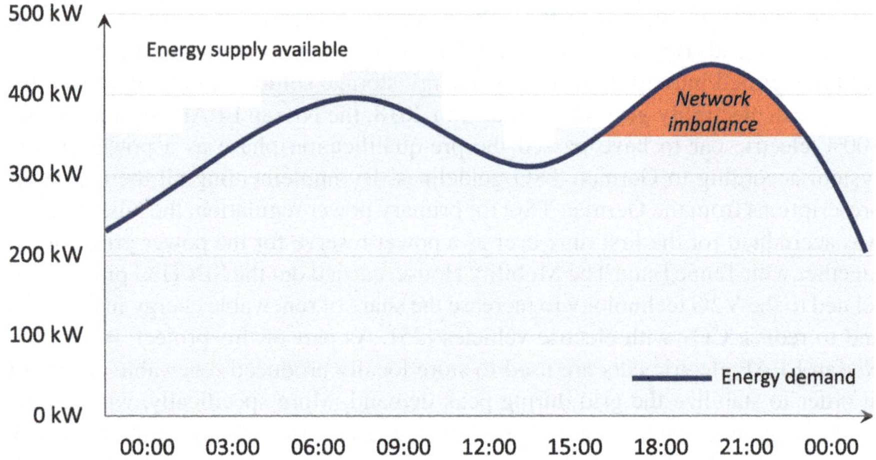

Fig. 3.5 Mismatch between supply and demand

balance generation and consumption [28]. The advantages of DR are broadly recognized and analyzed in the literature, because by appropriately shifting and reducing the peaks of power demand, the use of costly generation systems (in terms of marginal costs) would be no more necessary. Finally, the costs related to congestion would decrease, less investments would be needed for the transmission and distribution networks reinforcements, and sector competition in liberalized markets would be stimulated. Likewise, if a renewable generation system inputs electricity when loads are low, an increase of electrical load in some other time may gets away from the curtailment of such sources, thus lowering the related economic losses. Sometimes, DR is considered synonymous of energy efficiency as it modifies demand/production in a given time of the day to better match the generation. Even if DR programs often have an effect on energy efficiency, DR and energy efficiency need to be distinguished. Some consumption profiles of electricity can be shifted to a different time of the day; others, if avoided at a particular time, because no more needed, they will not be transferred to a different time. For example, if an end user chooses not to use a shiftable load [29] (like a washing machine) at a given time, he/she can use it in another time of the day, but if the same end user uses air conditioning (non-shiftable load) at less power, he/she will not maybe turn it on again, since it may not be needed in other times of the day. In this latter case, the use of electricity is reduced, but not delayed and this is the case of energy efficiency. DR is thus an important tool to avoid technical issues in the network and keeps the costs of the electricity system operation low. At the same time, DR allows the penetration of higher shares of renewable electricity in the grid and allows to avoid the need for peak power plants to cope with imbalances in the network.

DR can be in general implemented acting on two items: *capacity* and *balancing* [28].

The purpose of working on capacity is to limit peak electricity demands, so end-users participate in the program by making their own load capacity available. The peak demand occurs only a few times every year and can be reduced, without DR, only by putting up new generation capacity. As far as residential consumption is concerned, it is certainly possible to delay/anticipate some loads or modulate them [30], even though these operations may impact on end users' comfort. Furthermore, as DR operates based on the coordinated operation of many low energy consumption appliances, an aggregator is required to take part to the energy market. As a result, there would be no necessity to install new costly power generation plants to cope with peak loads and, on the other hand, end users would receive a remuneration for having met the DR requests [31, 32]. When considering the capacity feature of DR, the end-user (industrial or commercial/residential aggregates) can choose to turn off or shift some loads during a few hours per year in order to reduce the total load in the network during the period where a peak load is expected. The customer load reduction can thus be done at three levels:

1. industrial level;
2. commercial level;
3. residential level.

Considering the first level, some industrial end-users can easily vary the time for many of their processes, so they can shut down parts of the industrial process (shiftable-load), reducing their power consumption during the DR event, and to resume the process after the DR event. The commercial end-users can reduce the load for example by just dimming the lights of a lobby of a big hotel for a few minutes, providing stability to power system when demand increases. From the point of view of load management, the residential level is the one showing the easiest operation, because in residential units there are both shiftable and non-shiftable loads, therefore it is possible to shift some loads or to reduce them. In this way, there is no need to build new expensive power plants for facing the peaks and end-users can receive an economic benefits by meeting the DR requests [32].

When considering DR as balancing tool, the purpose is indeed to balance in an automatic way predictable variation in renewable energy output. This automated DR operates by connecting with a building's heating, ventilation, air conditioning, and lighting systems to adjust them in real time as grid conditions change. Using this type of DR program, network operators can manage direct regulation of the electricity consumption so as to face the rapid change in production from renewable energy generation plants. It is possible, as an example, to balance generation and consumption, thanks to the use of smart appliances (devices that can collect data from the grid operators about when to turn on or off). Another example may be the aggregation of electric vehicles to provide services for adjusting the frequency in real time when they are connected to the network. Consequently, using DR as means for balancing, it is possible to match the generation from renewable energy sources by tuning the demand in real time, to bring down energy price, as it remunerates the consumption from renewable sources and avoids renewable sources curtailment. Furthermore, it gives system operators a tool for implementing regulation and, finally,

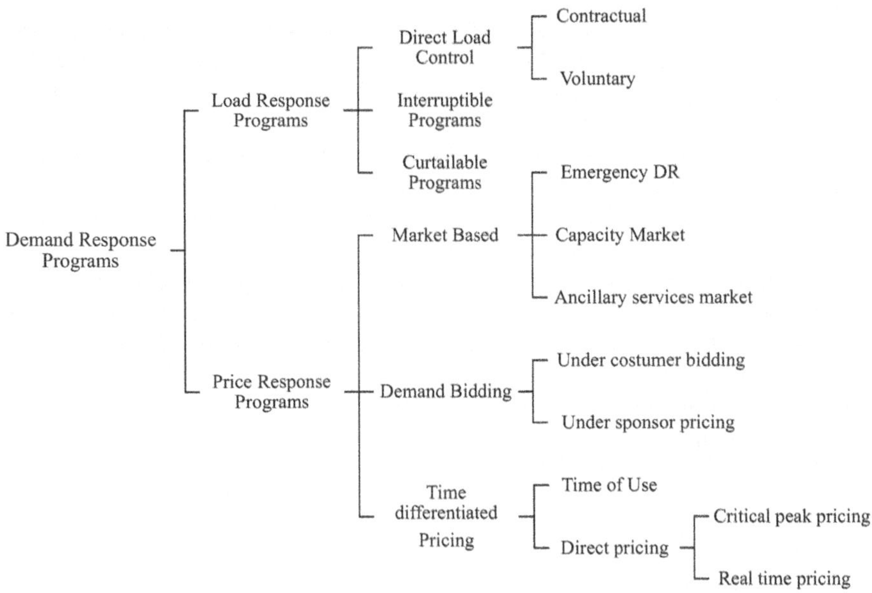

Fig. 3.6 Classification of DR programs

it creates a revenue stream for consumers. The final objectives of balancing and capacity based DR can be reached using different programs as depicted in Fig. 3.6 [29, 30, 33].

As depicted in Fig. 3.6, the DR programs can be split up into two main classes: *Load Response Programs* and *Price Response Programs*.

In the Load Response Programs, participants attain incentives or reduction in electricity bills by lowering consumption when the utility asks them to do so. The load lessening can be implemented by means of three different strategies:

1. *Direct Load Control*: the utility directly controls the end-user's load by using some remote control device in order to remotely control the consumer's appliances. This program is often used for residential customers and they can curtail their loads according to a contract (Contractual) or on a voluntary basis (Voluntary).
2. *Interruptible Programs*: the customer shuts down some loads for a short time interval. The participants can be commercial or industrial customers and receive upfront incentives or rate discounts, but they can face penalties if they don't respond to the program.
3. *Curtailable Programs*: the customers that cannot turn off their supply can participate to the DR event reducing their load.

The *Price Response Programs* is mostly used for industrial or commercial customers, which are offered an economic incentive for taking part in the program.

These programs can be divided in three categories: *Market Based*, *Demand Bidding* and *Time differentiated price*.

The *Market Based* programs focus on reducing the wholesale market price and can be of three kinds:

1. *Emergency DR*: customers who take part are remunerated for load lowering during emergency conditions.
2. *Capacity Market*: customers who take part can offer pre-defined load reductions as system contingencies occur.
3. *Ancillary services market*: participating customers can bid on load curtailment in the spot markets operating reserve.

In the second category, *Demand bidding*, the end-users make offers on specific load reductions in electricity wholesale market and can be divided in two branches:

1. *Under customer bidding*: customers bids the price that is lower than market price for lowering specified load at given time at most one day ahead or, in some cases, one hour ahead.
2. *Under sponsor pricing*: market administrators communicate the reduction price per kWh to the customers, who receive a reduced electricity price based on the amount of load reduction.

Finally, in the *Time differentiated pricing programs*, suppliers expose customers to variable electricity prices so that they can move loads from high-price periods to low-price periods with the consequent flattening of the load diagram. These programs are divided in:

1. *Time of use*: it is the simplest time based pricing program. The energy price is divided into two fixed blocks referring to different times and usually higher prices occur during peak hours and lower prices during off-peak hours. This program offers consumers the opportunity to reduce their electricity bills by shifting their load from peak hours to peak or valley hours.
2. *Direct pricing*: in this program the energy price follows the hourly variation of the wholesale market prices with two different mechanism:

    a. *Critical peak pricing*: used during contingencies or high wholesale energy prices. It is similar to the time of use program but it adds the third block called "critical peak period" and the blocks may not have a fixed price and a fixed time frame.
    b. *Real time pricing*: it follows the real time variation of energy price informing the end-user of price variation as little as of 5 min interval.

Another classification is based on how load changes are brought about. From this perspective, two categories are distinguished: *Price-based DR* and *Incentive-based DR*. The first refers to modifications in users habits as a response to changes of electricity tariffs. The second provides remuneration to end users for a modulation of the load; these incentives can either be paid separately or considered as a discount to retail electricity price. A number of DR programs inflict a penalty to end users who sign up but do not fulfil their contractual commitments when they are called for it [34].

3.4.2 Actors of the DR

As already said in paragraph 2.3.1, the aggregator is an actor of the electric market that has the task to supply capacity to the system operator grouping in virtual units the users of the electric system. The provision of this service is implemented by the execution of agreements, between the aggregator and these sets of virtual units, on the modulation of the power consumption both in reduction and in increase. During an event of DR the grid operator (typically the TSO) transfers the request for load modulation to the aggregator in a given area of the electrical grid. The aggregator, knowing the overall capacity of the end users of an area who have decided to take part to the DR service, responds to these orders from the grid operator by requesting, in compliance with the constraints for the aggregated resources, the load modulation to the end users that participate in the virtual load unit. By means of a public sale involving the grid operator and the aggregator, the power reduction/increase and the price that the grid operator will pay to the aggregator for the service are known and public. When the DR event ends, the aggregator pays the end users according to the designed DR program. The actors that operate in the DR programs deployment are different based on the type of market in which the same program acts. The following Table 3.1 illustrates the role of the main actors in the DR service, according to the responsibilities they take.

The role of the Aggregator can be played in different ways. In fact, the Aggregator can be an energy supplier that offers aggregation services, a service provider specialized in aggregation services that collaborates with a supplier or a joint venture between a traditional supplier and a service provider or an independent market player. Many different business models underlying aggregation are described in the literature, but not all of them can be actually implemented. Table 3.2 shows a summary of the different business models that are currently appearing throughout Europe, distinguishing between aggregators that combine different roles and independent aggregators.

Table 3.1 shows that the type of aggregator depends on the market phase in which it participates, the size of the production and consumption units it aggregates and the type of users it involves in DR programs. While Table 3.2 shows that the aggregator can perform only one or more activities depending on whether it is a combined or independent aggregator. In blockchain applications depending on the users involved and the tasks it performs, the aggregator can be both combined and independent. The type and business model may change from application to application.

Table 3.1 Main actors of the demand response service

Grid operator	Aggregator	Customers
TSO or DSO	***Parallel Aggregator*** participates in different market at the same time (electricity, water, heat)	– Industrial, commercial and domestic consumers – Big and small producers
	Large-scale Aggregator aggregates large production or consumption units connected in HV or VHV	– Industrial and commercial consumers – Big producers
	Micro Aggregator aggregates small production units	– Small producers
	Global Aggregator aggregates both small production and consumption units	– Small producers and consumers (i.e. domestic)
Role		
Notify the aggregator of the need for balancing on the network following the forecast of a stability problem	Identify the load and/or generation of its customers necessary to increase or decrease the use of energy. Remunerate customers whohave responded positively to the request	Implement the plans of modulation, according fto the DR programs discussed in the previous paragraph

3.4.3 The Baseline

The last stage of a DR program consists in the remuneration of the end users who have taken part to the program and have thus responded actively to the request to change the consumption as communicated by the Aggregator or, as in this case, by the DSO. In order to measure to what extent the end users have taken part to the program, and thus to assess the benefit they have brought by modulating their load, it is necessary the identification of a "Customer Baseline Load" (CBL). The CBL is the reference consumption pattern of each end user that takes part to the DR program and is used to assess the effects of the DR program implementation on a given end user or a set of end users' load profile. Indeed, the effect of the DR can be numerically assessed as the difference between the consumption during a DR event and the CBL (see Fig. 3.7). In the implementation of a DR program, the appraisal of the CBL is quite important, as the difference between the CBL and the real consumption is takes as the customers' performance under the DR program and is the reference to design the economic compensation mechanism. An accurate estimation of the CBL is critical to the accomplishment of DR programs as it engages the interests of many stakeholders including utilities and customers. In the state of the art, there are a number of methods to assess it [35–40], but not all of them can be used effectively in a DR program as they are too complex.

Table 3.2 Business models for aggregation

Business model	Description
Combined aggregator—Supplier	Supply and aggregation are offered as a package and there will be one BSP per connection point because aggregator and BRP are the same entity. The main benefits are reduced complexity and the absence of financial settlements between suppliers and aggregators
Combined aggregator—BSP	There are 2 BSPs on the same connection point, the BSP (independent aggregator) and the BSP (supplier). The aggregator has an agreement with the consumers of the supplier and the supplier will have to be compensated for the electricity that was sourced on day-ahead or other markets
Combined aggregator—DSO	DSO and aggregator coincide. Customers buyelectricity from retailers, but now the DSO provides DR services to them. DR results can be sold on the market as negative consumption. This can be done by the DSO or by a subsidiary to the DSO company due to regulatory issues
Independent aggregator as a service provider	The aggregator is a service provider for one ofthe other market actors but does not sell at own risk to potential buyers. The aggregator and the other market actor should have a long-term and exclusive relationship because, since the independent aggregator has no balancing responsibility, he will gainthe full benefits of his actions. The other market actor, nevertheless, is exposed to price risk
Independent delegated aggregator	The aggregator sells at own risk to potential buyers such as the TSO, the BSP and the wholesale electricity markets
Prosumer as aggregator	Large-scale prosumers, such as commercial and industrial prosumers,choose to adopt the role of aggregator for their own portfolios

The work in [39] also arranges CBL evaluation methods into three main classes:

1. *Averaging methods*: based on the hypothesis that the consumption profiles of an individual end user in several adjacent days are similar, these methods estimate the CBL based on the average consumption of days prior to the event day.
2. *Regression methods*: attempt to fit a linear function to describe the relationship between the consumption and some meaningful variables, such as historical load and weather data (e.g. temperature, humidity and wind speed) and then use this function to estimate the CBL of the event days.
3. *Machine Learning methods*: attempt to find the potential relation between the consumption and its impact factors. Differently from regression models, Machine Learning methods can find the hidden non-linear relation and exhibit high levels of estimation accuracy.

The methods for computing the CBL, particularly when these are parts of a shared business model, need to be transparent and easy to understand for both utilities and

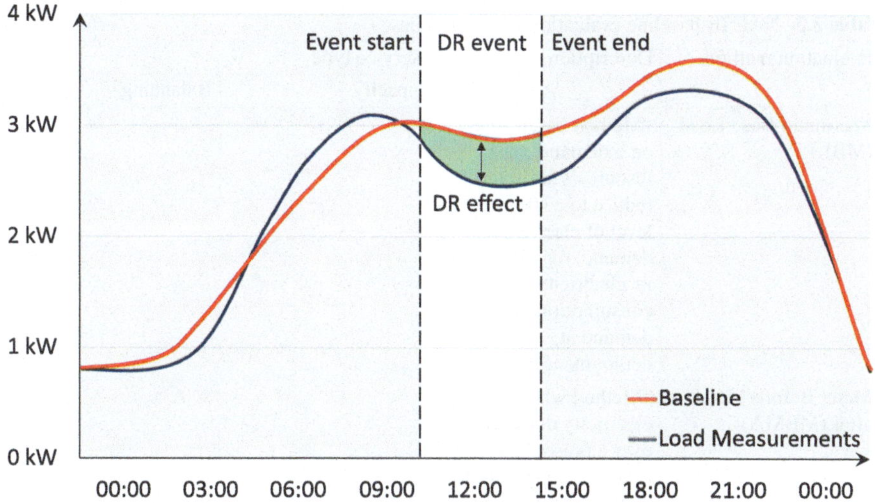

Fig. 3.7 Comparison of baseline with the actual load profile during the DR event

end users. Hence, although machine learning methods may deliver higher estimation accuracy, they are difficult to be applied in practice due to their inherent limited transparency. The increase in the number of methodologies for assessing the CBL increases the complexity of operating in the field of DR, making the choice of the methodology and the comparison to the others difficult for both customers, aggregators and BSP.

The North American Energy Standards Board (NAESB), an industry forum for the development and support of standards in wholesale and retail gas and electricity markets, developed a set of common definitions and practices in the area of DR. The NAESB has thus defined five kinds of CBL evaluation methods and has discussed the suitability of each method for each type of service provided by the DR program [41], see Table 3.3.

Each of these methodologies has numerous possible variations, but in general:

- *MBL* are adopted for verifying the contribution of demand resources to capacity commitments.
- *MBMA* are adopted for ancillary services (in particular frequency regulation).
- *BT-I* are the most commonly adopted, especially for measuring demand reductions offered in the energy markets.
- *MGO* are used when there are on-site generation units.

Depending on the type of DR program, it is possible to choose the CBL evaluation method, but in general the BT-I, like the averaging methods, are the most suitable for the blockchain applications because they are simple to apply, quite reliable and can be used for both capacity and balancing services.

Table 3.3 NAESB baseline evaluation methodologies

Evaluation method	Description	Service type	
		Capacity	Balancing
Maximum Base Load (MBL)	"Method based solely on a demand resource's ability to reduce to a specified level of electricity demand, regardless of its electricity consumption or demand at deployment"	x	
Meter Before/Meter after (MBMA)	"Method where electricity demand over a prescribed period of time prior to deployment is compared to similar readings during the sustained response period"	x	x
Baseline Type-I (BT-I)	"Method based on a demand resource's historical interval meter datawhich may also include other variables such as weather and calendar data"	x	
Baseline Type-II (BT-II)	"Method that uses statistical sampling to estimate the electricity consumption of an aggregated demand resource where interval metering is not available on the entire population"	x	x
Metering Generator Output(MGO) or Behind the-Meter Generation	"Method used when a generation asset is located behind the demand resource's revenue meter, in which the demand reduction value is based on the output of the generation asset"	x	

3.4.4 DR Benefits

Electrical systems have three important characteristics. Firstly, because electricity cannot be stored economically, the supply and demand must be kept in balance in real time. Secondly, grid conditions can change significantly from one day to the next or from one hour to the next, mainly due to the increasing penetration of energy produced by RES. Demand levels can also change quite quickly and unexpectedly, and as a result imbalances between supply and demand can jeopardise the integrity of the large scale grid in a matter of seconds. Thirdly, the electrical system is capital-intensive, as both large power plants and transmission systems have long delivery times and a multi-decade economy. These characteristics of electrical installations require that electrical networks are planned and operated years in advance to ensure that installations can operate efficiently in real time regardless the many doubts surrounding future demand. In the competitive electricity market, the organizations serving the load (integrated services or retail electricity suppliers) buy or build 60–95% of their offer in advance, with the belief that they will be capable of generating or buying enough electrical energy on the spot market in real time to follow the system's balance. With DR is possible to address these uncertainties by increasing the flexibility of the electricity system while keeping costs relatively low and facilitating the integration of renewable energy. Network operators, Independent System Operators (ISO), Regional Transmission Organisations (RTOs) or utilities and other organizations can employ the DR to reduce or shift loads instead of traditionally building more generation [42]. DR offers a number of financial and operational benefits to all actors of the electrical system. These advantages can be grouped into three main classes:

1. *Operation*: the electrical system requires large amounts of reserve generation to protect against generation fluctuations from RES. The energy generated by RES replaces the energy from conventional power producers, but the capacity of these generators needed to keep the security of the system is not enough. DR can provide these security services through load reduction and load shift. In addition, the use of DR to ensure system safety reduces the need to operate the power plants at partial load, which is inefficient and generally results in higher fuel costs. It also reduces dependence on the import and export of electricity through interconnections with neighbouring regions. This is particularly interesting from an economic point of view, as it allows these interregional links to be used only when it is profitable, rather than out of the need to balance the system.
2. *Planning*: a significant component of the total costs of the electrical system is represented by the costs of acquisition and maintenance of generation plants. The use of DR to reduce the generation capacity of the system could result in a substantial reduction in total costs. With DR it is possible to balance the fluctuations due to generation from RES thus avoiding the use of peak power plants, usually expensive. It also increases the usefulness of existing installations as more consistent production can be maintained. An additional benefit is the potential to reduce emissions from fossil fuel power generation. The DR service

is also useful to avoid grid congestion, thus maximizing the usefulness of the network and delaying or eliminating the need for upgrades and reinforcements of power lines.

3. *Economic*: the involvement of responsive demand in the electricity market has a number of key benefits. First, the power of market can be limited, allowing demand to respond to prices differently by geographic location and by duration, so that these features limit the ability of large generators to manipulate the wholesale price of electricity. In addition to the short-term efficiency gains related to prices, the DR has proved strong long-term efficiency gains in the form of efficient capacity planning. So, at the market level, the economic benefits consist in the lower wholesale market prices as DR prevents the need of using power plants with higher marginal cost to be operated in periods of otherwise high demand, lowering generation costs and prices for all purchasers of wholesale electricity. For end users, the economic benefits translate into savings on bills and incentives gained by adjusting the electricity demand in response to changes along time. In the longer term, a nurtured DR reduces the need for system's capacity, allowing the utilities and other retailers to purchase or build less new capacity.

The most important advantage of DR is undoubtedly the improvement of the efficiency of production resources due to a better balance between energy produced and consumed avoiding the use of inefficient and expensive peak power plants. The greatest potential for reducing peak demand and shifting over the day is held by residential users [43], rather than commercial or industrial users. But to date, in countries where DR service is used to address issues that can affect the electricity system, it is mostly large consumers who take part in the service. This work aims to provide a new DR model that can also be easily and efficiently applied to residential users.

3.4.5 State of Art

The issue of decentralized management of DR is discussed in many recent works that aim to propose new control and management models without the need for a central controller that exchanges data with all distributed resources. The study in [44] describes a blockchain platform for registering the energy usage data and energy generation data of the prosumers in a microgrid. Thanks to SCs executed on the Ethereum blockchain it is possible to assess the foreseeable flexibility of each prosumer, the related prize or penalty and the rules for the whole energy balance in the microgrid considered. The literature also points to some other approaches to the DR programs implementation based on the use of DLTs. Other interesting papers [45–47], suggest new methods to provide energy flexibility services to grid operators, both TSO or DSO, by means of DR programs implemented with blockchain platforms, but only accounting for the cooperation between grid operators and BSP not considering how prosumers implement the load changes. These papers suggest the integration of blockchain solutions in the existing DR programs with the objective of lowering information asymmetry for the end users.

DR programs for peak load and congestion management on power lines have been in use in several countries for many years, but its application has long been limited to large consumers. According to a recent report by the IEA [48], the global potential for technically available DR is about 4000 TWh and could reach 7000 TWh by 2040, with more than 85% of the volume growth attributable to the construction and transportation sector. Most of the potential is in the residential and commercial sector, involving air conditioning and household appliances. Also according to IEA, the full implementation of the technical potential would lead to about 200 GW of additional flexibility for the global electrical system in 2040. This level of DR would avoid investments in new electrical infrastructure (new power generation, transmission and distribution capacity) totalling 270 billion dollars. Despite growing interest, however, the DR market is still rather limited today (about 45 GW globally), partly due to existing regulatory constraints.

According to the European Commission, 20 GW of DR are currently used, but the potential in the EU could reach 160 GW in 2030. The rapid spread of renewables and the electrification of mobility and the civil sector will in fact increase the value of flexibility, thus increasing the economic value of DR interventions. DR in the residential sector is considered an emerging market in Europe, with an estimated 1.3 GW of activities already involved today [49]. The residential sector, in particular, has a great untapped potential of 200 GW, 97% of which is connected to air conditioning systems and water heaters. However, access to these resources for DR is challenging; less than 1% of all existing residential air conditioning and hot water systems currently have integrated connectivity.

To date, there are no real experiences but only projects in development such as DELTA [50], eDREAM [51] and BLORIN [52]. The H2020 DELTA project, proposes a platform consisting of virtual nodes to manage problems on the grid due to load peaks and imbalances through cooperation and interactivity of end users. In this situation, no SC manages the flexibility of the prosumers, rather a multi-agent system running on Virtual Nodes that updates the energy consumption/injection profiles of each prosumer in the aggregator's portfolio. The system is equipped with load forecasting and optimization procedures that provide the necessary information for self-balancing and prevent loss of stability and deterioration of quality of service. The European eDREAM project aims to develop innovative solutions for the implementation of real-time DR strategies supported by the use of blockchain technology. These technological solutions will allow an optimal and rapid detection of flexibility within the power grid, secure data management, microgrid control and new business opportunities. The eDREAM technologies will be accurately validated both in laboratory scale (Greece) and in two pilot sites (UK and Italy). The BLORIN project aims to create a technology platform based on the blockchain for the dissemination of renewable energy and the management of energy exchanges. This platform will help the creation and dissemination of solar Smart Communities and will be able to facilitate interactions between prosumers, manage charging infrastructure for electric vehicles and coordinate exchanges with the DSO. Blockchain technology will be used to manage energy exchanges between prosumers, to aggregate them and to provide DR services. The BloRin platform will be developed and tested at the Engineering Department of the University of Palermo, together with the industrial partners Exalto Energy & Innovation and Regalgrid, and will then be field tested in two Sicilian islands, through the contribution of distributors, S.EL.I.S. Lampedusa S.p.A. and SEA Societá Elettrica di Favignana S.p.A., active partners of the project. In Lampedusa a micro-grid involving a mix of photovoltaic systems and storage systems with the possibility to manage the demand profile of various users thanks to DR programs is being tested. In Favignana, the platform is employed for managing the electric vehicles' charging systems so as to improve their contribution as distributed storage for the improvement of the quality of grid operation.

These projects aim to respond to the challenges arising from the growing uncertainty in balancing energy due to the increasing contribution of renewable sources, proposing at the same time a useful tool for the evolution of the electricity market in a direction that involves end users on regional, national and international scenarios.

3.5 Suitability of the Blockchain for Power System Applications

As already discussed, the blockchain is an emerging technology for sharing decentralized data and transactions through a large network of untrusted participants. This new technology, thanks to the benefits it provides, has several use cases, but as a database and computing platform also has disadvantages as compared to conventional databases. The blockchain may be the right choice for some applications, while conventional technologies are more adequate for other use cases. One of the main issues to be addressed in first place is to decide whether the application of blockchain technology for power systems operation is suitable or not. A recent work about the adequacy of the application of the blockchain technology to a given practical problem provides a simple flowchart for this purpose [53]. The flow chart is reported in Fig. 3.8, with some adjustments introduced for blockchain applications to power systems use cases.

Considering the flowchart above, it is here tried to assess the real need of blockchain technology for the electricity sector. First of all, both in P2P electricity trading, such as in V2G applications and in DR related applications, the issue is certainly multi-party. Besides, smart meters and actuators at the prosumers' premises are decentralized sources of data, for which it is needed to have the whole transactions history especially for assessing offers and demands along time and for quantifying remunerations (e.g. in DR applications). Then, time constraints for blockchain can either be strict or not, as meters typically register energy in a given timeframe and it is always possible to extract synthetic indicators for any profile. Moreover, in these cases, there is a need for transparency from all the involved parties, generally to

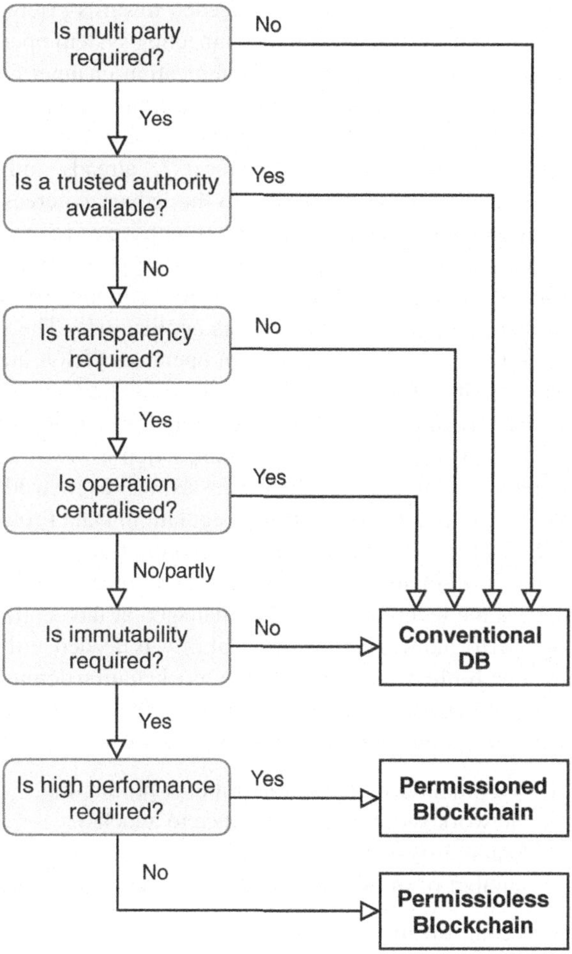

Fig. 3.8 Blockchain suitability evaluation graph for power system applications

overcome the traditional knowledge asymmetry between the system operator/energy retailers and end-users. Transparency is indeed a key feature especially for public and regulatory institutions, and it is generally desirable also for private companies. It is also worth to notice that often DSOs and TSOs are involved in legal disputes with aggregators, prosumers, consumers and energy retailers due to various reasons (theft of electricity, excessive voltage drops, loss of service continuity, damages to the grid due to faults at the user's installation, etc.). In many cases, however, thanks to the features of immutability, confidence and transparency, the blockchain could provide a quick solution to many disputes. This could modify the process described in [53] reinforcing the main role of transparency and immutability in the choice of the blockchain solution. Figure 3.8 shows that, without a trusted authority, if transparency or immutability are required, the use of blockchain is always suggested.

Full transparency, however, is needed towards pricing mechanisms and especially in the communication channel connecting system operator/aggregator/energy retailers and end-users. Such a communication channel can even be invisible or partly visible to other consumers. In this case, the requirements for performance can lead to different types of blockchain. The choice depends on the rules for access to the ledger and the permit to write over it. As already said, the blockchain can be public or private based on the access to the ledger, whereas it is named permissioned or permissionless depending on who is authorized to maintain the ledger. Furthermore, as compared to the proposal in [53], the flow chart is slightly modified including also the answer is not "partly", due to the many different features of cyber-physical problems, such as power systems applications. The power system sector involves several authorities such as system operators (TSOs and DSOs), aggregators, certification entities for physical apparatus providing data (IoT), Data Protection Officers (under GDPR 2018). Despite these authorities do not trust each other, the technical authorization for an energy transaction is always given by the system operator, while financial assets of the parties as well as their identities are guaranteed by the blockchain itself under national regulation. Data Protection Officers instead call for the need for a centralized administration of the data system, thus indicating permissioned blockchain infrastructures. Operation control can only partly be decentralized since some technical issues are managed at the central level (i.e. balancing). When high performance, in term of scalability, is needed, either security or decentralization must be limited, according to the blockchain trilemma assumption [54], thus again pointing at permissioned blockchain technologies. The blockchain trilemma claims that a blockchain system can provide at most two of the following three properties:

1. Maximal decentralization of block production.
2. Network security (its resistance to attacks).
3. Scalability of the network (its ability to support load growth, related to the number of transactions that the network can process in a given time interval).

Decentralization refers to a low computational burden for any node of the blockchain, meaning the possibility even for a low computational power unit to take fully part to the blockchain community. Security is the property for which even in the presence of malicious attacks, none of the contents of the blocks could be altered or other malicious actions are prevented (e.g. the double spending of utility tokens). Finally, Scalability is related to the possibility to increase the speed of transactions, namely the number of transactions that can be validated and agreed upon per second. This dimension concerns the efficiency of the technology in validating and "finding a consensus" on transactions contained in a block. Therefore, with respect to this item, both the technical validation if needed and the consensus algorithm have particular relevance.

Focusing on the three applications of energy blockchain detailed in this chapter and considering the flowchart proposed together with the features of the blockchain trilemma, it can be said that the blockchain technology, especially the permissioned one, is suitable for applications in the energy sector. Both in P2P, V2G and DR, considering the number of actors that could be involved, a permissioned blockchain is able to guarantee decentralization, security and scalability, without neglecting transparency and trust between network users.

Considering for example a microgrid, consisting of consumers and prosumers downstream of the same substation, a centralized system for implementing P2P energy exchanges between users does not fit with the decentralized architecture of the system. Moreover, the presence of a database implies the presence of an authority, which will have both decision-making and executive control over the transactions' execution, thus invalidating transparency. The same logic is valid for DR service and V2G applications that are naturally distributed across the territory. The blockchain is a technology born for distributed applications and that allows to overcome the problems that affected centralized systems when applied to distributed realities.

References

1. C. Long, J. Wu, C. Zhang, M. Cheng, A. Al-Wakeel, Feasibility of peerto-peer energy trading in low voltage electrical distribution networks. Energy Procedia **105**, 2227–2232 (2017). https://doi.org/10.1016/j.egypro.2017.03.632
2. NEMoGrid Consortium, *NEMoGrid*. Accessed 11 Sept 2020. [Online] Available https://nemogrid.eu/
3. M. Sabounchi, J. Wei, *A Decentralized p2p Electricity Market Model for Microgrids* (2018), pp. 1–5. https://doi.org/10.1109/PESGM.20.8586257
4. B. Neagu, G. Gheorghe, O. Ivanov, *An Efficient Peer-to-Peer bBased Blokchain Approach for Prosumers Energy Trading in Microgrids* (2019), pp. 1–4. https://doi.org/10.1109/MPS.2019.8759743
5. A. Alam, A. Ferdous, M. Islam, *Towards Blockchain-Based Electricity Trading System and Cyber Resilient Microgrids* (2019). https://doi.org/10.1109/ECACE.2019.8679442
6. R.K. Kodali, S. Yerroju, B.Y.K. Yogi, Blockchain based energy trading, in *TENCON 2018—2018 IEEE Region 10 Conference* (2018), pp. 1778–1783
7. S. Pee, E. Kang, J. Song, J. Jang, *Blockchain Based Smart Energy Trading Platform Using Smart Contract* (2019), pp. 322–325. https://doi.org/10.1109/ICAIIC.2019.8668978
8. E. Mengelkamp, J. Gärttner, K. Rock, S. Kessler, L. Orsini, C. Weinhardt, Designing microgrid energy markets: a case study: the Brooklyn microgrid. Appl. Energy **210**, 870–880 (2018). ISSN: 0306-2619. https://doi.org/10.1016/j.apenergy.2017.06.054. [Online]. Available http://www.sciencedirect.com/science/article/pii/S030626191730805X
9. L. Orsini, S. Kessler, J. Wei, H. Field, *10—How the Brooklyn Microgrid and Transactive Grid are Paving the Way to Next-Gen Energy mmarkets*, ed. by W. Su, A.Q. Huang (2019), pp. 223–239. https://doi.org/10.1016/B978-0-08-102207-8.00010-2. [Online]. Available http://www.sciencedirect.com/science/article/pii/B9780081022078000102
10. V. Junge, R. Rieckhoff, H. von Tschischwitz, A. Dombrowsk, *Enyway: It's Just a Revolution* (2019). Accessed 27 Aug 2020. [Online]. Available https://www.enyway.com/en/blockchain/whitepaper
11. C. Plaza, J. Gil, F. de Chezelles, K.A. Strang, Distributed solar selfconsumption and blockchain solar energy exchanges on the public grid within an energy community, in *2018 IEEE International Conference on Environment and Electrical Engineering and 2018 IEEE Industrial and Commercial Power Systems Europe (EEEIC / I CPS Europe)* (2018), pp. 1–4
12. Power Ledger, *Power Ledger White Paper* (2017). Accessed 27 Aug 2020. [Online]. Available https://www.powerledger.io/wp-content/uploads/2019/11/power-ledger-whitepaper.pdf
13. IBM Corporation, *Daring to be First, How to Pioneers are taking the Plunge into Blockchain* (2018). Accessed 16 Oct 2020. [Online]. Available https://www.ibm.com/downloads/cas/WN5RKW2D

14. F. Orecchini, *Vehicle to grid—come funziona* (2020). Accessed 27 July 2020. [Online]. Available https://www.fabioorecchini.it/vehicle-to-grid-come-funziona
15. B.K. Sovacool, J. Kester, L. Noel, G. Zarazua de Rubens, Actors, business models, and innovation activity systems for vehicle-to-grid (v2g) technology: a comprehensive review. Renew. Sustain. Energy Rev. **131**, 109963 (2020). ISSN: 1364-0321
16. S.M.C.V. Diaz, *Analysis of the Vehicle-to-Grid Technology: European Case Study Review*. Accessed 17 Oct 2020. [Online]. Available https://www.politesi.polimi.it/bitstream/10589/149693/2/Analysis%20of%20the%20Vehicle-to-grid%20technology.%20European%20case%20study%20review.pdf
17. M. Yilmaz, P.T. Krein, Review of benefits and challenges of vehicle-togrid technology, in *IEEE Energy Conversion Congress and Exposition (ECCE)*, vol. 2012, pp. 3082–3089 (2012)
18. EnelX, *Il veicolo elettrico e il v2g-v2h* (2018). Accessed 17 Oct 2020. [Online]. Available https://www.motus-e.org/wp-content/uploads/2020/09/Il-Veicolo-elettrico-e-il-V2G-V2H-1.pdf
19. Z. Su, Y. Wang, Q. Xu, M. Fei, Y. Tian, N. Zhang, A secure charging scheme for electric vehicles with smart communities in energy blockchain. IEEE Internet Things J. **6**(3), 4601–4613 (2019)
20. H. Liu, Y. Zhang, S. Zheng, Y. Li, Electric vehicle power trading mechanism based on blockchain and smart contract in v2g network. IEEE Access **7**, 160546–160558 (2019)
21. Enel, *Vehicle to Grid, l'auto diventa una batteria*. Accessed 25 July 2020. [Online]. Available https://corporate.enel.it/it/media/news/d/2019/05/vehicle-to-grid-auto-diventa-batteria-mobile
22. F. Orecchini, *FCA faráasorgere a Mirafiori la piúugrande infrastruttura V2G al mondo*. Accessed 25 July 2020. [Online]. Available https://www.fabioorecchini.it/category/energia-ambiente/
23. SSE Enterprise, *Bus2grid: London Bus Garage Becomes World's Largest Vehicle-to-Grid Site*. Accessed 21 Oct 2020. [Online]. Available bit.ly/Bus2Grid
24. Enel, *Energy on Wheels: V2g, Renewables and Grids*. Accessed 21 Oct 2020. [Online]. Available https://www.enel.com/media/explore/search-news/news/2016/08/v2g-innovation-renewables-grids
25. TSCNET Services, *Tennet Concludes Successful v2g Pilot*. Accessed 21 Oct 2020. [Online]. Available https://www.tscnet.eu/tennet-concludes-successful-v2g-pilot/
26. Energy Networks Association, *Smart Demand Response: A Discussion Paper* (2012). Accessed 28 Apr 2020. [Online]. Available http://www.energynetworks.org/modx/assets/files/news/publications/Smar
27. PJM, *Demand Response Strategy PJM Interconnection* (2017). Accessed 24 Apr 2020. [Online]. Available https://www.pjm.com/media/library/reports-notices/demand-response/20170628-pjm-emand-response-strategy.ashx
28. S. Aggarwal, J. Gu, *Two Kinds of Demand-Response* (2012). Accessed 25 Apr 2020. [Online]. Available https://energyinnovation.org/wp-content/uploads/2014/11/Two-Kinds-of-Demand-Response.pdf
29. X. He, L. Hancher, I. Azevedo, N. Keyaerts, L. Meeus, J.-M. Glacant, *Shift, Not Drift: Towards Active Demand Response and Beyond* (2013). Accessed 24 May 2020. [Online]. Available https://cadmus.eui.eu/bitstream/handle/1814/27662/FSR
30. M. Tahir, C. Haoyong, I. Idris, N. Larik, S. Adnan, Demand response programs significance, challenges and worldwide scope in maintaining power system stability. Int. J. Adv. Comput. Sci. Appl. **9**(6), 124–131 (2018)
31. D. Hurley, P. Peterson, M. Whited, *Two Kinds of Demand-Response* (2013). Accessed 21 May 2020. [Online]. Available bit.ly/DRasPowerSystemResource
32. P. Bradley, M. Leach, J. Torriti, A review of the costs and benefits of demand response for electricity in the UK. Energy Policy Spec. Sect. Trans. Pathways Low Carbon Econ. **52**(01), 312–327 (2013). https://doi.org/10.1016/j.enpol.2012.09.039
33. M. Albadi, E. El-Saadany, A summary of demand response in electricity markets. Electric Power Syst. Res. **78**, 1989–1996 (2008). https://doi.org/10.1016/j.epsr.2008.04.002

34. U.S. Department of Energy, *Benefits of Demand Response in Electricity Markets and Recommendation for Achieving them* (2006). Accessed 23 May 2020. [Online]. Available https://www.raponline.org/wp-content/uploads/2016/05/doe-benefitsofdrandrecommendations-2006-02.pdf
35. C. Lake, *PJM Empirical Analysis of Demand Response Baseline Methods Prepared for the PJM Markets Implementation Committee* (2011). Accessed 1 May 2020. [Online]. Available https://www.pjm.com/-/media/markets-ops/demand-response/pjm-analysis-of-dr-baseline-methods-full-report.ashx?la=en
36. L. Hatton, P. Charpentier, E. Matzner-Lober, Statistical estimation of the residential baseline. IEEE Trans. Power Syst. **31**(3), 1752–1759 (2016). https://doi.org/10.1109/TPWRS.20152453889. ISSN: 08858950
37. S. Mohajeryami, M. Doostan, A. Asadinejad, P. Schwarz, Error analysis of customer baseline load (CBL) calculation methods for residential customers. IEEE Trans. Ind. Appl. **53**(1), 5–14 (2017). https://doi.org/10.1109/TIA.2016.2613985. ISSN: 00939994
38. T. K. Wijaya, M. Vasirani, K. Aberer, When bias matters: An economic assessment of demand response baselines for residential customers. IEEE Trans. Smart Grid **5**(4), 1755–1763 (2014). https://doi.org/10.1109/TSG.2014.2309053. ISSN: 1949-3061
39. K. Li, B. Wang, Z. Wang, F. Wang, Z. Mi, Z. Zhen, A baseline load estimation approach for residential customer based on load pattern clustering. Energy Procedia **142**(12), 2042–2049 (2017)
40. T. Song,Y. Li, X. P. Zhang, J. Li, C. Wu, Q. Wu, B. Wang,A cluster-based baseline load calculation approach for individual industrial and commercial customer. Energies **12**(1), 1–17 (2019). https://doi.org/10.3390/en12010064. ISSN: 19961073
41. E. Winkler, P. Wattles, D. Pratt, *Measurement and Verification Standards Wholesale Electric Demand Response, Recommendation Summary* (2008). Accessed 9 Apr 2020. [Online]. Available https://www.naesb.org//pdf3/dsmee100308w7.pdf
42. N. O'Connell, P. Pinson, H. Madsen, M. O'Malley, Benefits and challenges of electrical demand response: a critical review. Renew. Sustain. Energy Rev. **39**, 686–699 (2014). https://doi.org/10.1016/j.rser.2014.07.098
43. O. Corradi, H. Ochsenfeld, H. Madsen, P. Pinson, Controlling electricity consumption by forecasting its response to varying prices. IEEE Trans. Power Syst. **28**(1), 421–429 (2013)
44. C. Pop, T. Cioara, M. Antal, I. Anghel, I. Salomie, M. Bertoncini, Blockchain based decentralized management of demand response programs in smart energy grids. Sensors **18**(162), 1–21 (2018)
45. A.C. Tsolakis, I. Moschos, K. Votis, D. Ioannidis, T. Dimitrios, P. Pandey, S. Katsikas, E. Kotsakis, R. Garcia-Castro, A secured and trusted demand response system based on blockchain technologies, in *Presented at 2018 Innovations in Intelligent Systems and Applications*, Thessaloniki, Greece, 3–5 July 2018. https://doi.org/10.1109/INISTA.2018.8466303
46. E. Riva Sanseverino, M. Ippolito, G. Zizzo, S. Favuzza, D. Arnone, M. Mammina, E. Telaretti, Demand project: Bottom-up aggregation of prosumers in distribution networks, in *Presented at 2018 AEIT International Annual Conference*, Bari, Italy, 3–5 Oct 2018. https://doi.org/10.23919/AEIT.2018.8577336
47. N. Liu, X. Yu, C. Wang, C. Li, L. Ma, J. Lei, Energy-Sharing model with price-based demand response for microgrids of peer-to-peer prosumers. IEEE Trans. Power Syst. **32**(5), 3569–3583 (2017). https://doi.org/10.1109/TPWRS.2017.2649558. ISSN: 08858950
48. IEA, *Demand Response* (2020). Accessed 11 Nov 2020. [Online]. Available https://www.iea.org/reports/demand-response
49. Future Grid Solutions, *Residential Demand Response: Pilot to Profitability* (2020). Accessed 11 Nov 2020. [Online]. Available http://futuregridsolutions.co.uk/residential-demand-response-pilot-to-profitability/
50. Delta Consortium, *Delta*. Accessed 28 Aug 2020. [Online]. Available https://www.delta-h2020.eu/
51. eDREAM Consortium, *eDREAM*. Accessed 29 Aug 2020. [Online]. Available https://edream-h2020.eu/

52. BloRin Consortium, *BloRin*. Accessed 29 Aug 2020. [Online]. Available https://www.blorin.energy/
53. X. Xu, I. Weber, M. Staples, Design process for applications on blockchain, in *Architecture for Blockchain Applications* (Springer International Publishing, Cham, 2019), pp. 93–111. https://doi.org/10.1007/978-3-030-03035-3_6
54. Q. Kaihua, G. Arthur, *An Overview of Blockchain Scalability, Interoperability and Sustainability*. Accessed 22 Oct 2020. [Online]. Available bit.ly/AnOverviewOfBlockchainScalability

Chapter 4
Development of a Blockchain Network for DR Service Traceability and Certification

G. Sciumè, E. Riva Sanseverino, P. Gallo, M. L. Di Silvestre, and G. Zizzo

Abstract This chapter aims at describing the integration of a blockchain platform for the execution of DR events into an existing power network without the need of additional devices and using the SMs already installed. For this purpose, the following key aspects are described: the current protocol used by many States for the execution of DR events, the blockchain platform used, the developed architecture and the logic implemented for the development of a new system for the execution and certification of DR events.

4.1 Introduction

DR programs are a set of activities with the aim of optimizing the consumption or production of electricity at specific hours of the day to meet RES penetration or peaks of demand. These programs are generally implemented through a centralised communication architecture where data and operation logic over data are controlled by a single entity, for example by means of a Supervisory Control & Data Acquisition (SCADA) system. In current DR program implementations, an aggregator is responsible for sharing the load change request among the end users according to a determined criterion and based on data that are not visible to end users. End-users are expected to be increasingly involved in DR programs due to the highly stochastic behavior of production units currently being connected to the grid, which will require a greater flexibility of energy resources in the system and therefore also including the end-users. This variability has suddenly increased the need for balancing services and near real-time markets. These activities can be managed by appropriate technology that transparently manages communications between end users and aggregators with limited delay, when the latter's demand for modulated power adequately accounts for the high variability of supply and demand.

G. Sciumè (✉) · E. Riva Sanseverino · P. Gallo · M. L. Di Silvestre · G. Zizzo
University of Palermo, Viale delle Scienze, Building 9, 90128 Palermo, Italy
e-mail: giuseppe.sciume01@unipa.it

G. Sciumè et al. (eds.), *A Practical Guide to Trading and Tracing for the Energy Blockchain*, UNIPA Springer Series, https://doi.org/10.1007/978-3-030-96607-2_4

However, end users are barely involved in DR programs for ancillary services and, when this happens, they only provide some capacity to be activated by directly switching on/off the loads, such as in Finland and UK with heating electric loads [1, 2]. More often, end users may take part to secondary and tertiary regulation.

As already illustrated through the literature review above reported, DR programs can be implemented using DLTs such as blockchain. Although the use of the latter technology for DR services provision entails transparency, information symmetry and security to participants to the platform, it also leads to some challenges that are detailed below.

First of all, the European General Data Protection Regulation (GDPR) calls for the existence of one liable party and one entity, the Data Controller, natural or legal person, public authority, agency or another entity, which, alone or together with others, identifies the aims and the means for processing personal data. The compliance with GDPR with respect to this problem entails the use of *permissioned* blockchain platforms. Blockchain platforms are indeed currently deployed as *permissioned* and *permissionless*. As explained in Chap. 1, *permissioned* blockchain needs prior approval from one entity before taking part to it, whereas *permissionless* blockchain lets anyone participate to the system. In *permissioned* blockchain, the identity of participants is known, while participants are anonymous in *permissionless* blockchain.

Another issue in using the blockchain for DR certification relates to transaction implementation times. The *permissionless* blockchain guarantees greater security, but, due to the computation time of the underlying mining algorithms, it is not adequate for managing operations close to real time. Another important feature is the huge energy consumption, which makes its application in the process of decarbonization of the energy sector not adequate. For this reason, a more suitable choice concerns the use of a *permissioned* blockchain for energy-related applications, such as DR programs implementation.

The Third Energy Package of the European Union [3], earlier in 2009, asked all Member States to provide for the installation of intelligent metering systems for the long-térm benefit of consumers. Such metering systems are usually called Smart Meters (SMs). With these devices, end users should also be able to access dynamic electricity price contracts. The Commission report *Benchmarking Smart Metering Deployment in the EU-27 with a focus on electricity* indicates precise targets for an extensive deployment of SMs: 72% of end-users in EU-27 were expected to have SMs by 2020. However, only some countries "precisely Sweden, Finland, Italy, Estonia, Malta, Spain and Denmark" have already a wide-scale roll-out for electricity smart meters in place today. As in most member states, in Italy, the implementation is deployed by the Distribution System Operators (DSO). The European Parliament directive on common rules for the internal electricity market identifies two drivers. The first one is to allow DSOs to operate local flexibility resources. In this way, network operation costs could be significantly reduced. Another key driver for competition and end users engagement is information. Earlier Commission consultations and studies have demonstrated that consumers complain of a lack of transparency in electricity markets, which lowers their opportunity to benefit from competition and to actively take part to the markets. Consumers do not enjoy sufficient information about different suppliers or the availability of new possible energy services and business models. They also criticize the complexity of offers and procedures for changing suppliers. The regulation also guarantees data protection as the growing use of new technologies (especially smart metering systems) generates an abundance of energy data, with high commercial value [4]. However, current SMs are not "smart" enough from the point of view of a distributed infrastructure. The "smartness" of the metering infrastructure is essentially connected to the ability for a DSO to read metered data remotely and removing the need for customer consumption estimates. In addition, they can provide metered data directly to customers through a dedicated local channel.

SMs for the Italian electricity market are a good example. In their version 2.0 (SM2G), they have two separate communication channels. The first channel is directed to the central distribution system (*chain 1*) to exchange information with the DSO and validated data with retailers. The second channel (*chain 2*) is directed to any local domestic automation system for smarter management of home consumption and appliances, although it provides non-validated data. The *chain 1* accommodates real-time communication using A-band PLC technology, while the*chain 2* employs C-band PLC. The latter guarantees the same or even better performance as A-band PLC as it is subject to limited disturbances [5]. The lack of connectivity between Italian SMs and the Internet (*e.g.* IoT) and the inability to install client code on them for security reasons preclude the ability to use them as blockchain nodes. This is currently not possible and creates some issues when it comes to opening up the market to third parties that could manage the blockchain platform.

The same situation is found in other member states such as Denmark, whose SMs diffusion status has almost reached the 100%. The Danish smart metering infrastructure pursues a similar setting to the Italian one, with a local communication channel in which the end user can visualize the domestic appliances demand by means of displays or monitoring systems while an external channel sends data to the DSO's concentrator. In the case of Denmark, proprietary RF or Wireless M-bus standards are generally adopted between the SMs and the data concentrators. The local interface, instead, generally has a serial port where Internet-based communication modules can be connected. This grows the flexibility of the system as compared to the Italian case, but it is still not sufficient to directly support extensive blockchain implementations [6].

4.2 Blockchain Technology and OpenADR

Most of the DR programs are currently implemented as centralized systems: decisions on loads control are taken by the grid operator or aggregator, on the basis of dynamic programming optimization results [7], fuzzy logic-based decisions [8], or other profit maximization schemes [9, 10]. These decisions are then proposed to the customers. This is a result of information asymmetry and centralized control over the available information deduced by the smart metering devices. Furthermore, pricing strategies are defined in a centralized way, even if a decentralized dispatch, based on price signals, influencing the users' choices [11] is implemented. In [11], the decisions are taken in a distributed and cooperative way, but transaction validation and DR logic are still managed centralized. However, it will be demonstrated that the design choices that might appear as a limiting factor for the blockchain potential implementation, on the other hand, reproduce a realistic scenario, where all market actors are present and take part. In a nutshell, DR programs are largely founded on a classical client–server architecture, in which a server holds the connections with industrial, commercial or residential end users for collecting measurements and sends control signals. Figure 4.1 shows a general architecture where the System Operator is able to exchange information and DR signals with the customers through a client-server messaging infrastructure.

The most used protocol to exchange data and signals in contexts in which DR programs are implemented is the Open Automatic DR, OpenADR [12]. OpenADR is a protocol that is fully open access and is implemented on a client server system architecture. It uses HTTP protocol for the ISO/OSI transport layer implementation, while Public key cryptography ensures security. OpenADR was developed to allow a stable transmission of DR signals to a participating customer. This gives the latter a continuous visibility of wholesale prices and helps to better balance supply and demand. The advantages are to facilitate a timely and predictable response for the system operator, while considering the end user's preferences. OpenADR creates a common language, the ADR 2.0 protocol, to communicate a DR event over an

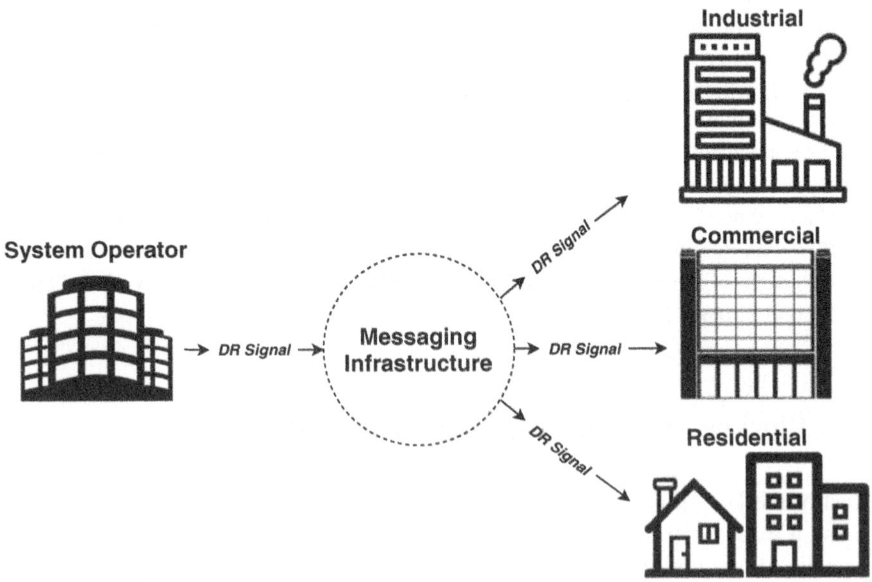

Fig. 4.1 General DR messaging architecture

IP-based network, such as the Internet. OpenADR 2.0 is the foundation of today's worldwide standard and is operated by the OpenADR Alliance. Many providers, located in several U.S. states, parts of Europe, China, Japan, Australia and Korea, now support this protocol, which ensures operation when a system operator calls an event. OpenADR works by having pre-programmed systems to take actions based on receiving a signal. System operators initiate a DR event through a Demand Response Automation Server (DRAS). The DRAS is responsible for communicating specific details about the event, such as duration, start time and price signals to the end-user devices. The DR signal communication is always carried out by using two actors, VTN (Virtual Top Node) and VEN (Virtual End Node). A VTN always controls many end nodes and is responsible for transmitting event specifications, such as price, and programming signals. DR aggregators, or those who sign up end customers and sell their reduced loads on the wholesale market, can be either the top node or a final node, depending on who is calling the event, see Fig. 4.2

The customers are the VENs providing the demand reduction to the aggregators. Examples of messages conveyed via OpenADR from VTN to VEN include:

- PRICE_ABSOLUTE: the price per kilowatt–hour.
- PRICE_RELATIVE: a change in the price per kilowatt–hour.
- PRICE_MULTIPLE: a multiple of a basic rate per kilowatt–hour.
- LOAD_AMOUNT: a fixed amount of load to shed or shift.
- LOAD_PERCENTAGE: the percentage of load to shed or shift.

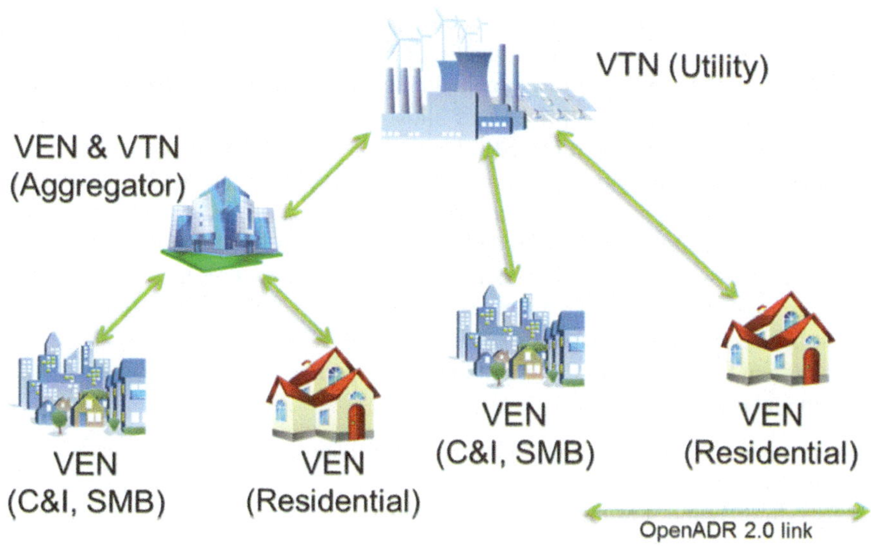

Fig. 4.2 OpenADR actuators architecture

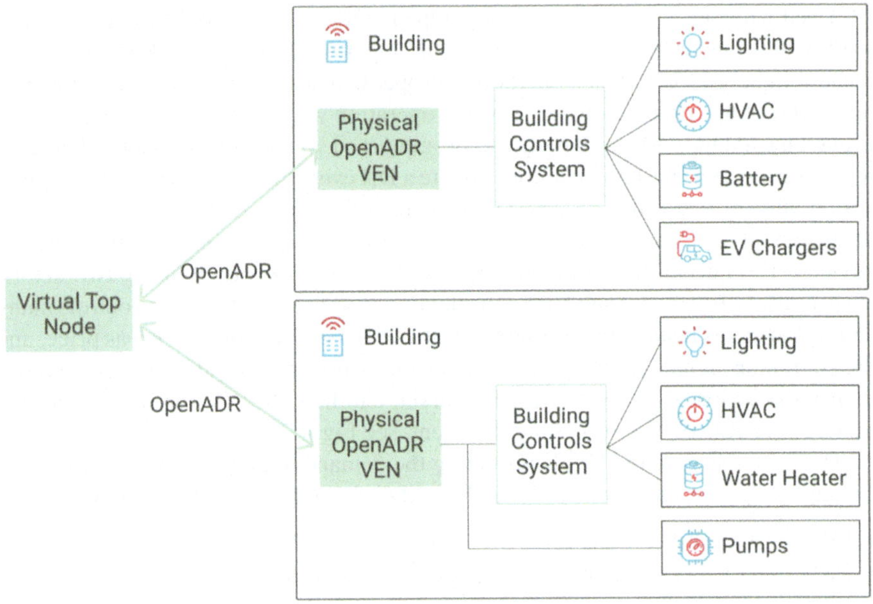

Fig. 4.3 VEN integration in buildings

A VEN can be integrated into the hardware of the on-site energy management system, or as a separate hardware device to pass signals directly to the end user devices or building control system, see Fig. 4.3.

The OpenADR protocol allows to obtain high speed in response from customers, give a constant visibility of prices and allow to choose the devices on which to act according to the signals received. The high speed in response allows to support ancillary services such as frequency regulation. This service can only be achieved by means of real-time signalling as it requires fast response, often in two to four seconds. Nevertheless, the architecture used, although efficient, suffers from the weaknesses of classic client-server systems, such as low scalability or the presence of centralized elements, since the failure of a VTN leads to the failure of the connected VENs and consequently the non-participation of users in the DR event.

The use of blockchain for DR as compared to OpenADR standard allows to obtain three main benefits:

1. Transparency over measures and pricing mechanisms.
2. High scalability.
3. No single point of failure.

Figure 4.4 shows a possible abstraction of the two communication models.

The OpenADR standard uses Hypertext Transfer Protocol (HTTP) as a transport layer. HTTP is ideal for pull clients and possible for push ones if security issues are handled. Also, the OpenADR standard uses Extensible Messaging and Presence

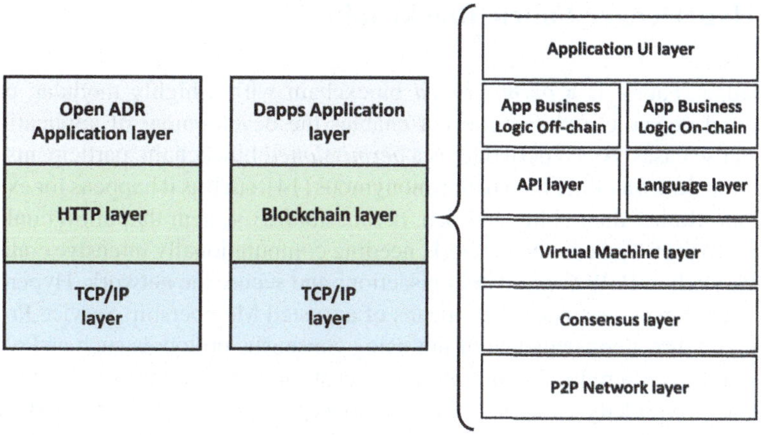

Fig. 4.4 Abstraction model for OpenADR and blockchain

Protocol (XMPP) as a transport protocol. This protocol is ideal for push applications and fast DR while pull is also possible. Virtual End Nodes may utilize HTTP or XMPP, while both of the protocols are mandatory for a VTN. Thus, the OpenADR standard event communication process is a push-pull action between the VTN and the VEN.

As it can be seen, blockchain technology, as transport layer of the distributed applications (aApp), is based on a number of layers: the P2P network layer is in charge for inter-node communication, discovery and data transfer (usually transactions and block propagation). The consensus Layer comprises pieces of code needed to generate the order of 'blocks creation' and to validate blocks created by other nodes in the network. The virtual machine layer is a "transactional engine", in charge for modifications in the blockchain's world state. The fourth blockchain abstraction layer consists of two parts: Application Programming Interface, API, employed by on-chain runtime applications, and Coding languages, used during development and compiled in runtime into a binary code that can be registered into blockchain and understood by the VMs. The language level relates to the SC development language. For Application business (on chain and off chain) layer, the code is generally written by third-party developers and not by core blockchain teams. This occurs because these are application-specific projects that use the underlying blockchain, so as to deliver some vertical solutions. The last layer is the User Interface, UI, of the dApp visible to the user [13].

The blockchain, compared with OpenADR, is more complex, but at the same time allows the user to obtain several advantages and overcome the weaknesses of OpenADR model, such as the integration of the platform with existing systems without the need to install third-party software. In this case, the VTN and VEN logics are implemented by the blockchain (for example through a SC) rather than by an external application.

4.3 Hyperledger Fabric Blockchain

Hyperledger Fabric is a *permissioned* blockchain with a highly modular, configurable, and versatile architecture that enables the development of applications in different use cases. As Hyperledger is a *permissioned* blockchain, participants to the platform are known rather than being anonymous [14], such as it happens for example in Bitcoin. Rather than being an open, *permissionless* system that allows unknown identities to participate in the network, needing computationally intensive consensus protocols such as PoW to validate transactions and secure the network, Hyperledger Fabric members are registered by means of a trusted Membership Service Provider (MSP). The direct consequence of not using consensus protocols such as PoW and mining process is a reduction in transaction costs and execution time.

Besides, Hyperledger Fabric is the first blockchain platform to support SC, called "chaincodes", written in generic programming languages like Java, Go and Node.js [15]. As introduced in Chap. 1, a SC is a piece of code that implements a shared logic for processing data and managing access and changes to a set of key-value pairs in the "current ledger state". This important feature makes this blockchain an easy-to-use tool for a wide variety of distributed applications. The Hyperledger Fabric blockchain is founded on a modular architecture that organizes in a logical way the different parts of the node, and then of the network, into different containers. Figure 4.5 shows the main modular components of the Hyperledger Fabric network.

In Fig. 4.5 the elementary network consisting of five parts is depicted: a Peer, a SC, a copy of the ledger, an Orderer and an App client. The status of the ledger is implemented as a local database. Hyperledger Fabric employs either LevelDB or CouchDB databases. LevelDB is the default database and is especially adequate if ledger states are simple key-value pairs. CouchDB is the appropriate choice if the ledger states are structured as JSON documents because CouchDB supports richer queries and updates of richer data types. The main network component is the peer as it can record the image of the ledger, it also hosts and runs the SC in order to implement

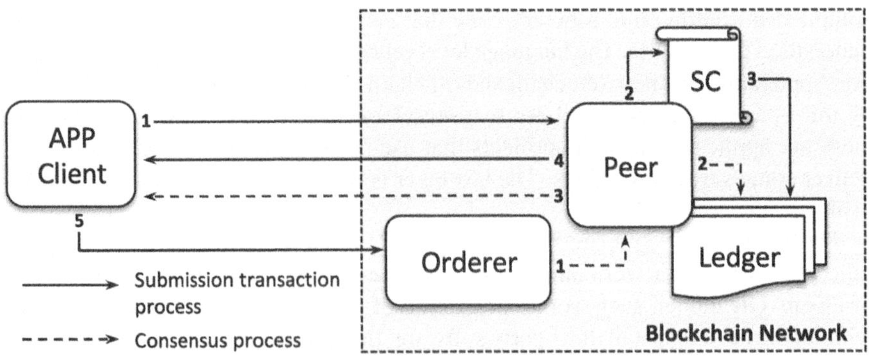

Fig. 4.5 Main components of an Hyperledger Fabric node

read/write operations in the ledger. Peers are owned and maintained by members of the blockchain network. Outside the blockchain network, the so called App client is needed to communicate with the peer. It also shows to the end-user or network's administrator the outcome following a query or a "transaction proposal". The solid line arrows of Fig. 4.5 show the process of execution and diffusion of a transaction that is illustrated here below. The network member, through the App client, can sign the "transaction proposal" and sends it (1) to all peers taking part to the blockchain. Each peer accepts the signed transaction proposal and invokes the SC (2) calling up a function that queries the ledger (3) and generates a "proposal response". The proposal response signed by all peers are sent to the App Client of the user who submitted the transaction proposal (4), that checks the signatures of all peers and compares the replies of the proposal to determine whether they are the same. If there is a match, the following step is taken. If the transaction proposal is a query, the latter is not sent to the Orderer. Rather it is sent and directly displayed by the App Client to the user who submitted the original transaction proposal. When the transaction proposal concerns a modification of the ledger status and after comparing the match of all responses obtained from the peers, the App Client transmits the response within a "transaction message" to the Orderer (5) which is the component responsible for the consensus process. The transaction message will contain the transaction data and the signatures of the peers. The Orderer does not have to inspect the precise contents of a transaction to carry out its operation, it simply receives the transactions, sorts them chronologically and generates the transaction blocks. Transactions are registered in the ledger in the order in which they occur, even if they involve different sets of participants in the blockchain. In order to let this happen, it is required to determine the arrangement of transactions and create a method to reject wrong transactions that were included in the ledger by mistake (or maliciously). In Hyperledger Fabric this task is entrusted to the Orderer. The dashed arrows in Fig. 4.5 show how the updating of the ledger is carried out after the consensus process performed by the Orderer. Once the order of transactions is identified, the new transactions are transmitted to the peers in blocks (1), the peers update the ledger with the new block of transactions (2), and in the end, the peers send a message (3) to the App Client to communicate the ledger update.

Another important feature of Hyperledger Fabric, compared to other blockchains, is the ability to choose the consensus protocol that best represents the existing relationships between participants. Up to version 1.0, there were two consensus mechanisms available, namely *SOLO* and *Kafka*. *SOLO* is the easiest mechanism, which just transmits the transaction without providing any real consensus. Clearly, it is not recommended for applications involving energy production. Differently, *Kafka* employs a fault-tolerant distributed streaming platform named *Apache Kafka* [16]. The latter also enacts distributed ordering service, and, as a consequence, it can have multiple Orderer nodes to prevent a unique point of failure [17]. In addition, Fabric's consensus protocols do not require a native cryptocurrency to foster expensive mining or SC execution. The platform can be deployed at the same computational cost as any other distributed system, due to the non-existence of heavy cryptographic mining. The composition of these characteristics makes Fabric a very high-performing platform in terms of transaction processing and transaction confirmation latency, and also guarantees privacy, transaction confidentiality, and SCs execution. With regard to privacy within the network, Hyperledger Fabric allows to select three different options. The first one consists in the use of communication mechanisms, called "channels", which allow the isolation and confidentiality of data exchanged. On a given channel, the sharing of the ledger is carried out among peers taking part to that channel, while transacting parties have to be properly authenticated in the channel in order to connect and interact each other. Hyperledger Fabric gives the ability to generate multiple channels in the same network, letting a group of participants to generate a separate transaction ledger. If two participants participate in a channel, only those two participants can have copies of the ledger for that channel, keeping the privacy and confidentiality of both of them. Figure 4.6 shows a dual-channel network in which Peer 1 and Peer 3 have access to only one ledger, while Peer 2 has access to both ledgers. All participants to the Hyperledger Fabric blockchain are usually clustered into organizations and more organizations can be clustered into consortia. An organization might host more than one peer and one client.

Starting from the version 1.2, and aiming to keep the confidentiality of data in the network beyond creating different channels, Fabric provides a second option, letting designers create "private data collections," which allow a given subset of organizations in a channel to support, commit, or query private data not needing to create a separate channel [18].

The last solution is to use attributes embedded in the digital identities of network members to determine permissions to use various functions implemented by the SC. The identities of all network participants are encapsulated in an X.509 digital certificate and are certified and verified by an MSP. Thus, an organization's MSP is required to verify the identities of users belonging to that organization. Even though the SC is the same for all peers and can be viewed by all participants to the blockchain, by the use of attributes it is possible to determine permissions to access the different functions implemented by the SC and handle privacy between different users in the

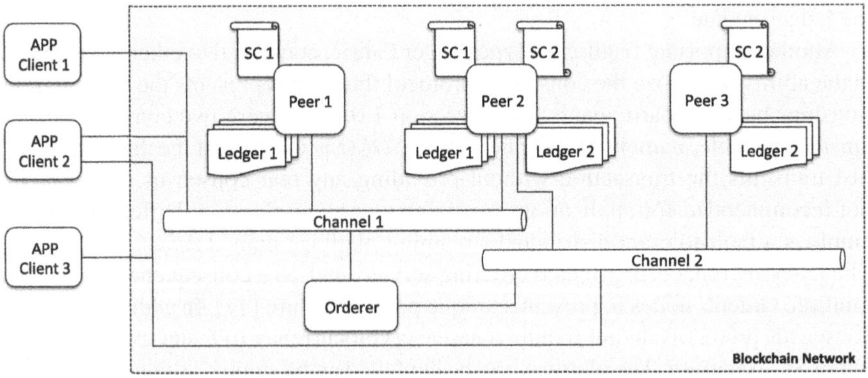

Fig. 4.6 Double channel Hyperledger Fabric network

network [19]. Adopting this solution, each user can read from the blockchain solely the data about him, such as his own baseline, consumption or earned tokens, but cannot read data regarding another user.

4.4 Integration of the Blockchain into Existing Power Grid Architecture

The proposed blockchain-based platform, developed using a Hyperledger Fabric network, was designed for DR programs implementation. The system was deployed taking into account the option of integrating with technologies that are already in place typically used for the low-voltage network. Considering an electrical microgrid composed of consumers and prosumers and taking into account the use of the Italian second-generation SMs, Fig. 4.7 shows the overall blockchain based architecture.

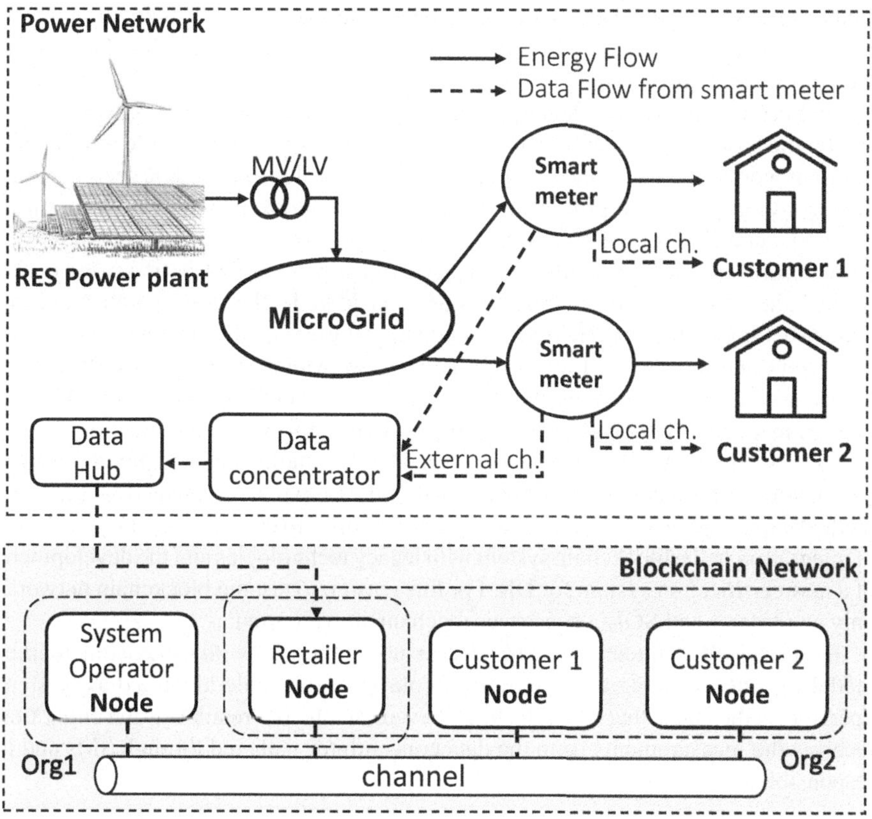

Fig. 4.7 Proposed blockchain-based architecture for DR [20]

In this architecture, consumption data are collected from the second (or external) channel of the SMs and are then transferred to the data concentrator. Then the same measures are transmitted by the retailer to the blockchain. The data concentrator can resemble a network's single point of failure, but, in general, this is not the case as several concentrators could be implemented to serve districts or local areas. Besides, in case of failure on the data concentrator, the users' consumption data will be registered by the SMs and transmitted to the blockchain by the retailer afterwards. The platform continues to operate even if a data concentrator fails. Concentrators require the computing power to act as clients of the blockchain and off-the-shelf embedded devices are adequate for this purpose. In addition, users can verify that the data taken from the data concentrator has not been tampered with by checking the match between the measurements provided by the retailer and those provided by the SM local channel. The aggregation device solves the problem of directly connecting the current SMs to the blockchain and provides a data stream that is trusted by the customer by allowing further validation by checking the correspondence with data from the retailer and DSO. In addition, customers have the ability to deploy their own client applications, reading data from the SM and checking if it matches the data recorded by the retailer on the blockchain. Users of the blockchain, through the client application, invoke a purpose-developed SC to process, in a distributed way, the transaction logic.

Furthermore the SC, invoked by the grid operator requesting the DR service, calculates the baseline for each user by reading the measures recorded in the blockchain. In a few words, the SC is the component to fully implement the DR service, from the recording of the load profiles of the users on the blockchain to the execution of the events and rewarding the users.

According to the 4.3 section, to manage roles, privacy and permissions in the blockchain network, two organizations are considered. The grid operator and the retailer belong to the first one (Org1), while the retailer and the customers belong to the second one (Org2) (see Fig. 4.7). Moreover, two Membership Service Providers, MSPs, are taken into account, one for Org1 and one for Org2. Thus, the MSP for Org1 is managed by the grid operator and is required to verify the identities of the resellers, while the MSP for Org2 is managed by the retailer and is required to verify the identities of the customers. Privacy within the network is managed by the third method explained in 4.3. The purpose of this section is to demonstrate the possibility of integration of the blockchain system with legacy technologies and the development of a new certification system for DR. For this reason, to test the blockchain network, only an Orderer and SOLO consensus mechanism were used.

The proposed architecture can be successfully integrated with the existing market model for DR by giving a role to existing players. The Node identified as 'system operator' is the entity that requests the DR event, while the 'retailer' is the entity that receives the measurements from the data concentrator gathered through SMs and is responsible for energy billing.

4.4.1 Smart Metering System

In the proposed blockchain based architecture, it is hypothesized that the SMs are owned and managed by the grid operator (typically the Distribution System Operator), while the billing is operated by the retailer. This one, here described, is the most frequent scenario among the EU Member States. The grid operator uses the acquired data both for operating the power grid (*e.g.* monitoring and controlling power losses and voltage drops) and giving retailers validated data to be employed for billing. Next, the grid operator gathers the measured data and makes them available to the retailer by means of the so-named "data management hub". This flow of information has the main aim to provide suppliers with validated consumption measures for billing. To make sure that collected data are sufficient and consistent, a validation process is carried out. This process is largely sufficient for the billing phase by means of advanced data reconstruction algorithms [5]. Figure 4.7 shows a community acting as a microgrid that is capable of exchanging energy with the grid. On the other hand, each end user has a different consumption pattern, as a result, an individual Smart Meter is required for each end user. A 3-phase Kamstrup OMNIA meter, very common in Denmark, was used for the testing phase. The load course of the end users were thus transmitted by the SMs to the data concentrator through an external channel of the SMs themselves. The protocol and technical standards for data exchange between the SMs and the data concentrator is EN 13757-5, implementing a radio mesh topology. Besides, the smart meters used in the tests are IEC 62056 compliant, which is the international standard of the DLMS/COSEM specification (Device Language Message Specification/Companion Specification for Energy Metering). The Smart Meters employed for the experimental application here described belong to the Kamstrup OMNIA suite and they are compliant with international standards. The electrical features they can measure are reported below: active positive energy (EN 50470-1 and EN 50470-3), reactive energy, and active negative energy (IEC 62052-11, IEC 62053-21 and IEC 62053-23), as well as Power Quality features according to the EN 50160 [21] standard. As the load data were available in the Data hub, the retailer could take them and subsequently record them over the blockchain. As a consequence, with this technology, all parties involved are sure that the recorded load data have not been altered by untrusted parties before being sent to the blockchain. Moreover, as a further guarantee, customers could verify the authenticity of the data using the local channel of the Smart Meter.

4.5 DR Event Formulation

The formalization of a DR event here considered, puts up a scenario in which the grid operator asks the end users to modify (increase/decrease in a given time frame) their load profiles to meet some technical requirements of the grid. All actuations are carried out through different functions of a Smart Contract that has been developed on purpose. The grid operator transmits DR signals to ask for a change of the loads profiles, in the range of the available loads flexibility attained as the summation of the flexible consumption powers declared by the customers. It is indeed clear that each customer has a different inclination (availability) to follow the suggested behaviour related to the DR signal. In detail, consumption behaviours are modelled to adapt according to an *availability function* that analytically represents the customer's willingness to take part to DR events. A set of customers are flexible enough to modify their typical load profile, others do not want or cannot adapt. Customers' classical consumption profiles are the so-named *baselines*, they are references to calculate the actual customers' adjustment to DR events. All customers that are part of the cluster are considered to be taking part to DR events. The latter are assumed to happen following the steps below, which are run periodically:

1. SMs register the customers' load consumption patterns and transmits them to the data concentrator.
2. The retailer records in the blockchain the load consumption pattern for each customer.
3. Ahead the communication of a new DR event, the grid operator prompts the baselines and availability profiles calculation functions of the Smart Contract.
4. The grid operator publishes the DR event in the blockchain, by writing the day, the time window and the total load reduction/increase.
5. The SC, taking into account the baselines and the availability profiles of the customers evaluated, distributes the overall load power reduction/increase request to the customers.
6. The day of the DR event, each customer reads from the blockchain the desired load reduction/increase evaluated by the SC so as to meet the grid operator's request.
7. The DR event takes place.
8. The customers' load consumption during the DR event is registered in the blockchain as explained in steps 1) and 2).
9. The day after the DR event, the market operator prompts the SC to verify the compliance of customers' consumption with the grid operator's request.
10. If the customers' consumption profile has been compliant with the request, the SC remunerates the customers with energy-tokens.

The baseline of a customer c is made of a vector of usual power consumption values along 24 h. The generic expression of such vector is given below in 4.1.

$$\mathbf{B}^{(c)} = \left[\bar{P}^{(c)}_{B,1}, \bar{P}^{(c)}_{B,2}, \ldots, \bar{P}^{(c)}_{B,h}, \ldots, \bar{P}^{(c)}_{B,24}\right] \tag{4.1}$$

Load values referring to a given hour h of the day d are dynamically averaged over more days. This averaging operation is indicated by the horizontal bar above the symbols of power consumption $\bar{P}^{(c)}_{B,h}$. In the following equations, the weekdays baselines and the weekends baselines are calculated.

$$\bar{P}_{B,h}^{(c)} = \frac{1}{X} \sum_{j \in \text{High}(X,Y,d)} P_{B,h,j}^{(c)} \quad \forall\, h \in \{1, 2, \ldots, 24\} \tag{4.2}$$

for the weekdays baseline, and

$$\bar{P}_{B,h}^{(c)} = \frac{1}{X} \sum_{j \in \text{Low}(X,Y,d)} P_{B,h,j}^{(c)} \quad \forall\, h \in \{1, 2, \ldots, 24\} \tag{4.3}$$

for the weekend baseline.

The term on the left side in 4.2 and 4.3 is the baseline value at hour h. This value is attained by calculating the average the power of the X days with the highest consumption 4.2, or the lowest consumption 4.3, in the Y non-DR days before the day d, namely the day in which the DR event is announced to customers. Day d is not specified, as it is the day in which the DR event is announced to customers, which is considered to be the day ahead the DR event occurs. Both for weekdays and for weekends, the methods based on averaging use the same set of Y days that do not comprise DR or other curtailment programs. The baselines are assessed over Y days ahead, in which the *HighXofY* method has been applied for the weekdays baseline, and the *LowXofY* method for the weekend baseline. In the end, the sliding window over the last Y days allows to take into account seasonality. The typical consumption profile of aggregated loads is the vector containing the sum of all the baselines for the customers that are part of a cluster CL, as expressed in 4.4.

$$\mathbf{B}^{(CL)} = \sum_{c \in CL} \mathbf{B}^{(c)} \tag{4.4}$$

As the grid operator sends a load reduction request for a determined hour h, the client application at the customer's premises executes the Smart Contract to verify what is the assigned load reduction in hour h leading to the largest economic compensation. This load reduction assigned to each customer is computed by the Smart Contract considering the respective "*Availability Profile*" (A). The latter is the customer's aptitude to meet a DR request from the grid operator. This figure of quality represents customers' performance and is composed of two terms. The first, named *difficulty*, gives information on how difficult it is to actually produce a load reduction as requested by the grid operator. The second term tells how good is the customer's compliance while meeting a DR request, called *compliance*. Besides, $\boldsymbol{A}$ changes its value at any new DR event. The superscript (c) has been removed in all the formulas below, to improve clarity. The availability profile $\boldsymbol{A}$ of a generic customer is a vector of 24 entries: as many as the hours in a day. These 24 elements of the vector $\boldsymbol{A}$ numerically represent the availability to modify the load at a given hour h, during the *i-th* DR event. This numerical value is made of two terms that are weighted by β and $1 - \beta$, both ranging between 0 and 1, as in 4.5.

$$A^*_{h,i} = \beta \frac{\Delta P_{B,h,i}}{\bar{P}_{B,h}} + (1-\beta) \frac{\sigma}{\sigma + \Delta P_{DR,h,i}} \tag{4.5}$$

In this formula, $\Delta P_{B,h,i}$ is the difference between the baseline $\bar{P}_{B,h}$ at hour h and the target consumption $P_{t,h,i}$ at the same hour; σ is the sensitivity of the SM (typical value is taken as 30 W, considering an accuracy of 1). $\Delta P_{DR,h,i}$ is the absolute value of the difference between the customer's load profile measured during the DR event and the desired (target) load profile. The two terms of the *Availability Profile* are both normalized as $\Delta P_{B,h,i}$ / $\bar{P}_{B,h} \in (0, 1]$ and $\sigma/(\sigma + \Delta P_{DR,h,i}) \in (0, 1]$. These two terms are computed at the *i-th* DR event occurring at hour h. The first term considers what fraction of the baseline is required to change, $\Delta P_{B,h,i}=|\bar{P}_{B,h}\text{-}P_{t,h,i}|$ (See Fig. 4.8). The higher is the first term, the larger the effort requested to the customer during the DR event.

The actual "capacity" of the customer is accounted for by the second term, $\Delta P_{DR,h,i} = |P_{h,i} - P_{t,h,i}|$. The "capacity" is defined as the difference between the measured load profile $P_{h,i}$ and the target load $P_{t,h,i}$ profile (deriving from the reduction requested by the grid operator) for the customer c at the hour h. This term is the customer's capability to "follow" the directions or requests from the grid operator. The smaller the $\Delta P_{DR,h,i}$, the more the customer load profile is compliant to the system operator's request. In 4.5 $\Delta P_{DR,h,i}$ is compared to σ, which is the sensitivity of the SM (accuracy class times full-scale of the SM) so when $\Delta P_{DR,h,i}$ is close to zero, the second term is close to 1. In 4.5, the value of $\beta \in [0, 1]$ defines which of the two terms is the most significant to assess the customer's behaviour. While β higher than 0.5 rewards those customers who were required more effort in the adaptation

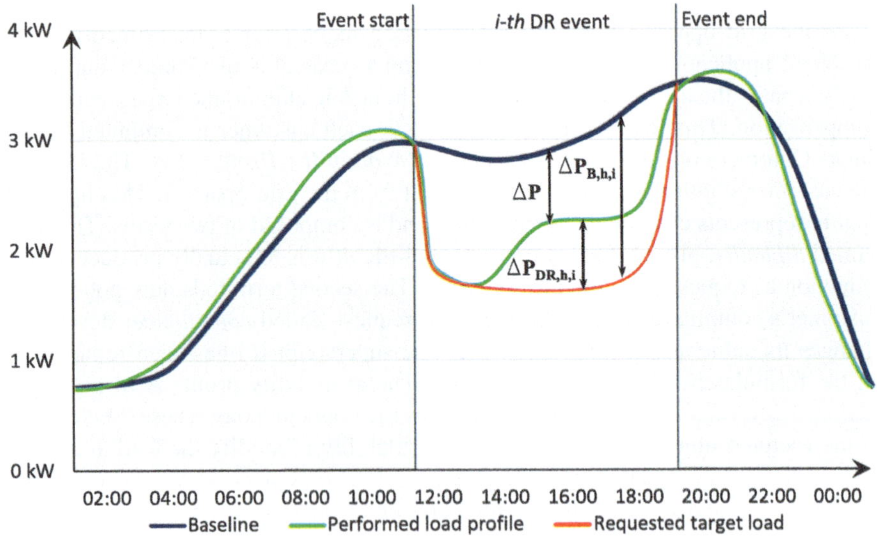

Fig. 4.8 Quantities involved in an exemplary DR event

(first term), even if they have low compliance (second term). β lower than 0.5 instead, rewards customers that pursue accurately the target profile (second term), even if this action does not require a big effort (first term). It is here heuristically suggested to set $\beta = 0.2$ as the second term (compliance) has a greater effect on stability rather than the first related to difficulty. Indeed, in the long term, the system is expected to require greater effort to those that acted correctly and, as a consequence, the difficulty for them is expected to increase in subsequent DR events. However, as the required effort increases too much, the customer is probably expected not to be able to respond correctly. As a result, the availability profile would reduce in a closed-loop self adjustment. Furthermore, if the numerical evaluation of the difficulty is dominant ($\beta > 0.5$), a greater reduction is to be assigned to those that were requested more effort, without taking care about their ability to follow the requests. In Fig. 4.9 it is shown how the availability profile depends on $\Delta P_{B,h,i}$, and $\Delta P_{DR,h,i}$, the courses are hypothesized to be referred to domestic users whose residential committed power is 3 kW. The same figure describes how the $A_{h,i}$ values, reported in the z-axis, range between zero and one in the different cases outlined above for $\beta \in \{0.2, 0.5, 0.8\}$.

Figure 4.8 shows a condition in which, from 11 a.m. to 4 p.m., the customer did not fully meet the reduction asked by the grid operator ($\Delta P_{DR,h,i} \neq 0$), while he fully satisfied the request from 16:00 to 19:00 ($\Delta P_{DR,h,i} = 0$). As a result, if the difficulty to meet the request is high for the customer, even if he was not fully compliant, the

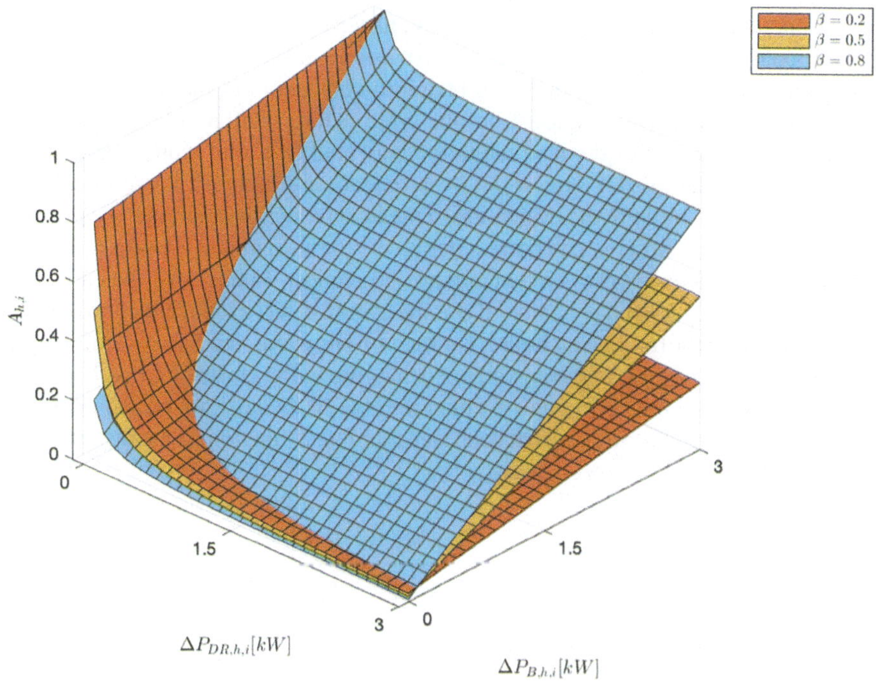

Fig. 4.9 Availability of a customer at hour h for the DR event i, when $\beta \in \{0.2, 0.5, 0.8\}$

availability will stay high anyway. If the difficulty to meet the request is low, being not compliant is more critically considered and the availability is low. Thus both terms are then weighted using the factor β. The *Availability* is thus again computed for each DR event, but to consider past events and filter out outliers behavioural variations, a memory term is accounted for. The value of *Availability* gets increasingly precise along time as the number of DR events to which the user participates increases, employing the learning rate α. Such a real parameter gets a numerical value below 1. Consequently, the *Availability* of a generic customer evaluated after the *i-th* DR event can be expressed as follows:

$$A_{h,i} = \alpha \cdot A^{*}_{h,i} + (1-\alpha) \cdot A_{h,i-1} \tag{4.6}$$

where:

- i indicates the last DR event;
- $\alpha \in [0,1]$ is the learning rate for the time-based adjustment of the *Availability Profile* of the customer.

In this case, α has been given the value 0.2, to filter out abrupt behavioural changes of the users. As cleared out in the following sections, the Availability is employed to compute the fraction of load reduction requested to each customer that is part of a cluster *CL* (will be considered only a load reduction and not an increase to simplify the explanation). The simplest way to make the correct association of a load reduction to a customer belonging to a cluster is to divide such reduction in a proportional way using the ratio (baseline)/(baseline of the cluster) at a given hour. This methodology, however, does not take into consideration the availability of a customer to meet a given request from the grid operator. Differently, changing the rule of proportionality of the load reduction may lead to a total request that can be above or below the overall request. To overcome these problems, it was decided to consider the availability in the evaluation of the load reduction fraction to be given to each customer, and moreover, to correct iteratively the assignment in order to achieve a good correspondence between assignment and request. In this way, the entire DR policy gains reliability and precision. If ϕ is the overall desired load reduction notified by the grid operator for a given hour *h* for the *i+1* DR event, the Smart Contract can share this overall request to the customers belonging to the cluster *CL*. Such sharing operation is carried out taking into account the $A_{h,i}$ of the customers, and using the following iterative algorithm that at each step *j* assigns a fraction of the residual load reduction, $\Psi^{(c)}_{h,j}$, as described below:

$$\Psi^{(c)}_{h,j} = \phi_j \cdot \frac{(1 + A_{h,i})}{2} \cdot \frac{\bar{P}^{(c)}_{B,h}}{P^{(CL_j)}_{B,h}} \tag{4.7}$$

where $c \in CL_j$ and $CL_j \subseteq CL_{j-i} \subseteq \cdots \subseteq CL_0 = CL$:

$$\phi_j = \begin{cases} \phi & \text{if } j = 0, \\ \phi - \sum_{c \in CL_j} \sum_{k=1}^{j-1} \Psi_{h,k}^{(c)} & \text{otherwise.} \end{cases} \tag{4.8}$$

The entire load reduction requested to a customer at step j is, thus, described by:

$$\Delta P_{B,h,j}^{(c)} = \sum_{k=1}^{j} \Psi_{h,k}^{(c)} \tag{4.9}$$

In 4.7, the iterative algorithm, at each step, assigns to the cluster a fraction of the residual load reduction ranging between 0.5 and 1, as $A_{h,i} \in [0, 1]$. At any step j of the algorithm, the cluster CL_j changes. Infact, those customers that have already been required to put their loads to zero do not take part to the subsequent assignment steps, therefore $CL_j = \{c \mid \Delta P_{B,h,j}^{(c)} < \bar{P}_{B,h}^{(c)}\}$. The ending condition of the iterative algorithm is that the residual reduction is below a given threshold. As outlined in 4.5, if the customer is reliable and/or has a very high potential to meet the request, then, he/she will be asked a stronger effort along the subsequent DR event. The *Availability* is initially set to 0.5, while the user has not yet taken part to any DR event at a given hour h, having a neutral behavior. Consequently, at any DR event, the *Availability* will be updated according to 4.6 and the Smart Contract assigns a load change according to 4.7. In 4.7 the variation of both the individual and cluster baselines are useful to update the assignment of the load reduction to the customers, following their availability.

4.5.1 Reward Mechanism

Generally, customers taking part to DR programs, are expected to reduce or increase their load as described in Sect. 3.3.1. What changes among DR programs is the way in which customers respond to grid operators' requests and especially how they are remunerated for the service provided. In the management of DR events, currently, the grid operator sends the Aggregator or Balancing Service Provider the request for load modulation in a given portion of the electrical system. A contract with the Aggregator rules the participation of customers in DR programs. The contract sets an incentive for the customer, if the latter is able to meet the request, and a penalty when the request is ignored. Having evidence of the total available capacity of the customers in a given area who take part in the DR service, the Aggregator responds to the orders from the grid operator. The Aggregator requests then, in compliance with the constraints and requirements needed for the aggregated resources, the modification of the consumption of the end users that make up the virtual load unit [22]. By means of an auction between the grid operator and the Aggregator, the power reduction/increase and the price that the grid operator would recognize to the Aggregator for the service offered by DR, is predefined. After the DR event, the Aggregator prizes the customers of the virtual unit according to the running DR program [23]. It is evident that this system, as conceived now, is lacking in transparency towards the end users and the grid operator as well. With the proposed blockchain-based architecture, the Aggregation in virtual units of consumers/producers is carried out through the blockchain. As a result, the grid operator can transfer the DR request directly to customers. At the end of the DR event, the Smart Contract remunerates the customers based on the load variation provided in response to the request. On the day in which the DR event takes place, the customer's consumption is recorded by the Smart Meter and it is written by the retailer on the blockchain. At the end of the DR day, the Smart Contract gives each customer a remuneration based on the load adaptation with respect the baseline following DR's request. Another important feature of DR programs is customers' remuneration. Some of the remuneration methods described in the literature are arranged so as to recognize a constant reward for each unit of energy consumption, modified by the customer. Other methods express remuneration as a function of the variation of the customer's profile [22], of the time of the day (as in Time-of-use schemes) or of the kind of customer. In [24], the authors propose different remuneration schemes for five classes of customers with different features. In [25] the authors suggest a remuneration scheme that sets an incentive that depends on the customer's power changes following a DR program, the differences between the amount of power purchased by the utility on the Spot market and the retail price for customers. In this way, the loss for the Utility was limited. In the application here presented, a remuneration scheme based on the change of power consumption as compared to the baseline using a quadratic remuneration function is proposed and used. The latter scheme is typically adopted for economical evaluations in the electrical energy field [26]. For the DR event, it is expressed as follows:

$$r = \sum_{h=1}^{24} \mathcal{H}(\bar{P}_{B,h} - P_h) \cdot (\bar{P}_{B,h} - P_h)^2 \tag{4.10}$$

In this equation, $\mathcal{H}$ is the Heaviside function. the latter is 1 when the argument is positive, and 0 otherwise, thus letting remunerate only load reductions and not load increases. In 4.10, $\bar{P}_{B,h}$ is the value of the baseline at hour h, while P_h is the load profile measured at hour h and their difference measures the customer response during the DR event. In the considered scenario, the grid operator requests load reductions and only those that reduce their load will receive a remuneration, according to a parabolic law. Moreover, the proposed approach may be changed using different remuneration functions (i.e. based on the season, the day of the week of the DR event, etc.)

4.6 DR Program Implementation Through a Purpose Developed SC

In the considered architecture, the blockchain platform interfaces with the grid through the Smart Meters. It was assumed that the power grid is a microgrid supplying consumers, while the blockchain network is made up of a set of nodes. One node is owned by the system operator, one by the retailer and two are owned by the customers (see Fig. 4.7). As already explained, the blockchain carries out the aggregation of the customers in virtual units. In this way, the system operator can interact directly with the customers, sending the request of load adaptation in order to avoid a technical problem, like a congestion on the network, in given hours of the day. The developed application runs through the Smart Contract a capacity/incentive-based DR program applied to residential consumers and implemented using a Demand Response program. It was thus assumed that customers take part to the program by making their load capacity available. The experimental scenarios have been designed taking into account that customers may choose to shift some loads in response to a DR event, so as to reduce the total load when the peak load is expected. When the grid operator notifies a DR event, the Smart Contract running on the peer nodes of the blockchain network distributes the overall load reduction to the customers proportionally to their own baseline and their availability profile according with 4.9. Thus the customers meet the request by shifting some loads during the DR event time, so as to receive the maximum remuneration. The Smart Contract represents the most important feature of the developed blockchain system, as it is the network component that allows to implement the logic of any type of transaction, to implement the above-mentioned DR program and validate the integrity of each transaction sent by the network users. The Smart Contract, such as the data that reside in the blockchain ledger, is installed on each peer node. Each time a user submits a transaction proposal over the blockchain network, this proposal is processes by the Smart Contract of all peers. Then, the App Client of the customer who submitted the transaction receives the responses to the transaction proposal from the Smart Contracts of all the other peer nodes and compares the match of the responses obtained. If the responses received are consistent, it means that the submitted transaction is compliant with the logic implemented by the SC and that all peers have verified this consistency. The App Client can thus send the response within a transaction message to the Orderer node which inserts it into a new block and sends it out to all peers which update the ledger. Using the Smart Contract, a distributed consensus on transactions consistency is then achieved. In this application, the Smart Contract was implemented, using the general programming language "Golang" (Go). All transactions are processed by invoking properly structured Smart Contracts functions. The Listing below shows the Smart Contract entry point for the main functions call. In the following Chap. 5, it is detailed how these functions are employed and invoked by the App Client of the various network users in order to implement a DR event.

Listing 4.1 Main DR SC functions.

```
func (t*SmartContract) Invoke(stub
    shim.ChaincodeStubInterface)pb.Response{
    function, args:=stub.GetFunctionAndParameters()
    fmt.Println("invoke␣is␣running"
    +function)

    if function=="recordEnergy"{
        return t.recordEnergy(stub, args)

    }else if function == "queryEnergyValue"{
        return t.queryEnergyValue(stub, args)

    }else if function == "queryDailyEnergy"{
        return t.queryDailyEnergy(stub, args)

    }else if function=="evaluateBaseline"{
        return t.evaluateBaseline(stub, args)

    }else if function=="evaluateAvailability"{
        return t.evaluateAvailability(stub, args)

    }else if function=="notifyDRevent"{
        return t.notifyDRevent(stub, args)

    }else if function=="queryLoadReduction"{
        return t.queryLoadReduction(stub, args)

    }else if function=="getDRToken"{
        return t.getDRToken(stub, args)

    }else if function=="queryEarnedToken"{
        return t.queryEarnedToken(stub, args)
    }

    fmt.Println("invoke␣did␣not␣find␣func:" + function)
    return shim.Error("Received␣unknown␣function␣invocation")
}
```

References

1. P. Bertoldi, P. Zancanella, B. Boza-Kiss, *Demand Response Status in EU Member States* (2019). Accessed 20 Oct 2020. [Online]. Available bit.ly/DRStatus
2. Smart Energy Demand Coalition (SEDC), *Explicit Demand Response in Europe, Mapping the Markets 2017* (2017). Accessed 24 Oct 2020. [Online]. Available https://www.smarten.eu/wp-content/uploads/2017/04/SEDC-Explicit-Demand-Response-in-Europe-Mapping-the-Markets-2017.pdf
3. European Union, *Directive of 2009/72/EC of the European Parliament and of the Council of 13 July 2009 Concerning Common Rules for the Internal Market in Electricity and Repeal-*

ing Directive 2003/54/EC (2019). Accessed 23 Oct 2020. [Online]. Available https://eur-lex.europa.eu/LexUriServ/LexUriServ.do?uri=OJ:L:2009:211:0055:0093
4. G. Erbach, *Common Rules for the Internal Electricity Market* (2019). Accessed 22 Oct 2020. [Online]. Available https://www.europarl.europa.eu/RegData/etudes/BRIE/2017/595924/EPR
5. A. Pitì, G. Verticale, C. Rottondi, A. Capone, L. Lo Schiavo, The role of smart meters in enabling real-time energy services for households: the Italian case. Energies **10**(2), 1–25 (2017). ISSN: 19961073. https://doi.org/10.3390/en10020199
6. S. Zhou, M.A. Brown, Smart meter deployment in Europe: A comparative case study on the impacts of national policy schemes. J. Cleaner Prod. **144**(12), 22–32 (2017). ISSN 0959-6526. https://doi.org/10.1016/j.jclepro.2016.12.031
7. Y.-Y. Hsu, C.-C. Su, Dispatch of direct load control using dynamic programming. Trans. Power Syst. **6**(3), 1056–1061 (1991)
8. R. Bayindir, E. Hossain, S. Vadi, The path of the smart grid—the new and improved power grid, in *Presented at the International Smart Grid Workshop and Certificate Program (ISGWCP)*, Istanbul, Turkey, 21–25 Mar 2016. https://doi.org/10.1109/ISGWCP.2016.7548270
9. K. Bhattacharyya, M.L. Crow, A fuzzy logic based approach to direct load control. IEEE Trans. Power Syst. **11**(2), 708–714 (1996). ISSN: 08858950. https://doi.org/10.1109/59.496143
10. G. Sheblé, K.-H. Ng, Direct load control—a profit-based load management using linear programming. IEEE Trans. Power Syst. **13**(2), 688–695 (1998)
11. A.H. Mohsenian-Rad, A. Leon-Garcia, Optimal residential load control with price prediction in real-time electricity pricing environments. IEEE Trans. Smart Grid **1**(2), 120–133 (2010). ISSN: 19493053. https://doi.org/10.1109/TSG.2010.2055903
12. A. Yassine, Implementation challenges of automatic demand response for households in smart grids, in *Presented at 3rd International Conference on Renewable Energies for Developing Countries (REDEC)*, Zouk Mosbeh, Lebanon, 13–15 July 2016. https://doi.org/10.1109/REDEC.2016.7577546, no. 08
13. S. Pay, *Towards Ccommon Blockchain Architecture—An 'ISO OSI' for Blockchain Primer* (2017). Accessed 27 Apr 2020. [Online]. Available https://medium.com/@scanpayasia/towards-common-blockchain-architecture-an-iso-osi-for-blockchain-primer-778db4e5b35c
14. Z. Ma, J.D. Billanes, B.N. Jørgensen, Aggregation potentials for buildings-business models of demand response and virtual power plants. Energies **10**(10), 1–19 (2017). [Online]. Available https://ideas.repec.org/a/gam/jeners/v10y2017i10p1646-d115763.html
15. E. Androulaki, C. Cachin, C. Ferris, C. Murthy, B. Nguyen, S. Muralidharan, M. Sethi, C. Stathakopoulou, Hyperledger fabric: a distributed operating system for permissioned blockchains, in *Presented at the Thirteenth EuroSys Conference 2018*, Porto, Portugal, 23–26 Apr 2018, no. 30
16. H. Wu, S. Zhihao, K. Wolter, Performance prediction for the Apache Kafka messaging system, in *Presented at IEEE 21st International Conference on High Performance Computing and Communications; IEEE 17th International Conference on Smart City; IEEE 5th International Conference on Data Science and Systems (HPCC/SmartCity/DSS)*, Zhangjiajie, China, 10–12 Aug, pp. 154–161 (2019)
17. The Linux Foundation, *Hyperledger Architecture, Volume 1: Introduction to Hyperledger Business Blockchain Design Philosophy and Consensus* (2001). Accessed 20 May 2020. [Online]. Available https://www.hyperledger.org/wp-content/uploads/2017/08/Hyperledger
18. F. Benhamouda, S. Halevi, T. Halevi, Supporting private data on hyperledger fabric with secure multiparty computation, in *Presented at 2018 IEEE International Conference on Cloud Engineering (IC2E)*, Orlando FL, USA, 17–20 Apr, pp. 357–363 (2018). https://doi.org/10.1147/JRD.2019.2913621
19. A. Vazratkar, *Attribute Based Access Control in Hyperledger Fabric* (2019). Accessed 21 May 2020. [Online]. Available https://bit.ly/ABACinHyperledgerfabric
20. G. Sciumè, E.J. Palacios-Garcí, P. Gallo, E.R. Sanseverino, J.C. Vasquez, J.M. Guerrero, Demand response service certification and customer baseline evaluation using blockchain technology. IEEE Access **8**, 139313–139331 (2020). https://doi.org/10.1109/ACCESS.2020.3012781

21. E.J. Palacios-Garcia, E. Rodriguez-Diaz, A. Anvari-Moghaddam, M. Savaghebi, J.C. Vasquez, J.M. Guerrero, A. Moreno-Munoz, Using smart meters data for energy management operations and power quality monitoring in a microgrid, *Presented at IEEE International Symposium on Industrial Electronics (ISIE)*, Edinburgh, UK, 19–21 June 2017, pp. 1725–1731
22. P. Faria, J. Spınola, Z. Vale, Methods for aggregation and remuneration of distributed energy resources. Appl. Sci. **8**, 1283 (2018). https://doi.org/10.3390/app8081283
23. M. Tahir, C. Haoyong, I. Idris, N. Larik, S. Adnan, Demand response programs significance, challenges and worldwide scope in maintaining power system stability. Int. J. Adv. Comput. Sci. Appl. **9**(6), 124–131 (2018)
24. P. Faria, J. Spínola, Z.Vale, Aggregation and remuneration of electricity consumers and producers for the definition of demand-response programs. IEEE Trans. Ind. Inf. **12**(3), 952–961 (2016). ISSN: 15513203. https://doi.org/10.1109/TII.2016.2541542
25. D. H. Vu, K.M. Muttaqi, A.P. Agalgaonkar, A. Bouzerdoum, Customer reward-based demand response program to improve demand elasticity and minimise financial risk during price spikes. IET Gener. Transm. Distrib. **12**(15), 3764–3771 (2018). ISSN: 17518687. https://doi.org/10.1049/iet-gtd.2017.2037
26. D.C. Idoniboyeobu, S.L. Braide, Analysis of economic load dispatch, for generation expansion planning. Int. J. Sci. Eng. Res. **8**(2), 875–881 (2017)

Chapter 5
Experimental Tests

G. Sciumè, E. Riva Sanseverino, P. Gallo, F. Montana, S. Longo, E. J. Palacios-García, J. C. Vasquez, and J. M. Guerrero

Abstract This chapter presents the outcomes of the experimental implementation of the architecture described in Chap. 4 running a hardware testbed for the emulation of domestic electrical load profiles. Besides it shows how this experimental scenario integrates both with the legacy metering architecture and with the blockchain platform. Finally, the integrated use with an Energy Management System designed on purpose is also described with the aim of ensuring intervention and automating the end user's response to a DR event.

5.1 Experimental Setup

The proposed solution has been validated by means of a hardware testbed assembled on purpose for emulating the domestic load appliances and the behaviour of residential consumers. Two Households equipped with both shiftable and non-shiftable loads are part of the considered experimental setup. Besides, the two households are interfaced with the electrical grid by means of Smart Meters that register the aggregated power. These experiments were carried out at the Microgrid laboratory of the Department of Energy Technology at Aalborg University, Denmark [1].

The schema of the setup is shown in Fig. 5.1. It comprises four bi-directional DC/AC converters. The latter operate in grid-connected mode and their power references can be fully controlled. The real-time control platform dSPACE shown on the top of the setup is in charge of generating pulses for the converters and for real-time control. In this way, the primary control can be implemented supported by a Phase Locked Loop, PLL, and power control loops for each inverter independently. Besides,

G. Sciumè (✉) · E. Riva Sanseverino · P. Gallo · F. Montana · S. Longo
University of Palermo, Viale delle Scienze, Building 9, 90128 Palermo, Italy
e-mail: giuseppe.sciume01@unipa.it

E. J. Palacios-García
University of Leuven, Kasteelpark Arenberg 10, 3001 Leuven, Belgium

J. C. Vasquez · J. M. Guerrero
University of Aalborg, Pontoppidanstræde 111, 9220 Aalborg, Denmark

G. Sciumè et al. (eds.), *A Practical Guide to Trading and Tracing for the Energy Blockchain*, UNIPA Springer Series, https://doi.org/10.1007/978-3-030-96607-2_5

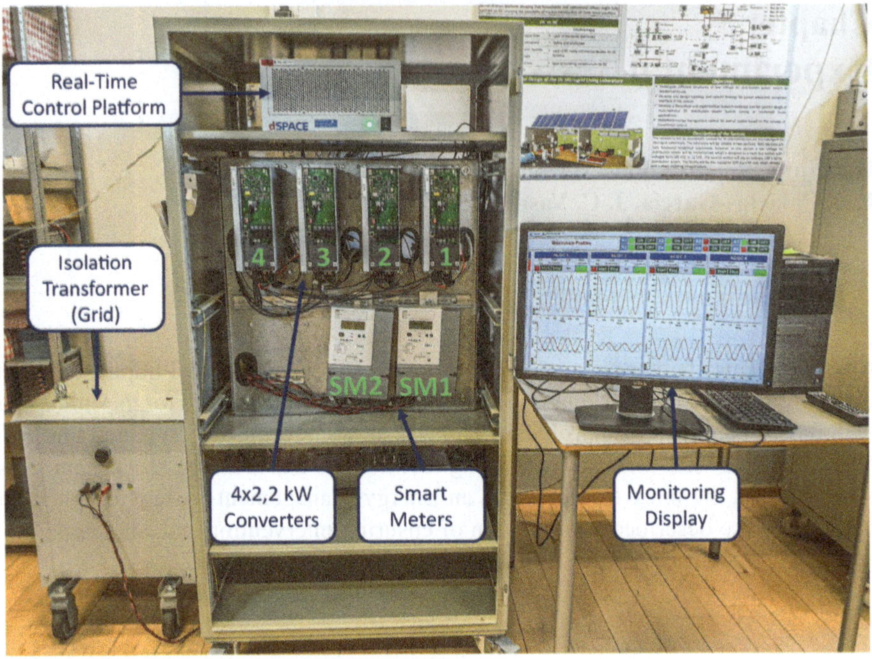

Fig. 5.1 Experimental test-bed emulating two residential consumers [1]

in the setup, it is also possible to have wireless control of the power reference set points by means of an MQTT interface. The same communication channel offers local monitoring functions both for voltage and current waveforms. MATLAB/Simulink has been used to program the dSPACE platform. Primary control loops, PLLs and power control loops were implemented and executed in MATLAB/Simulink, before being implemented in the system. The associated monitoring is carried out by means of the ControlDesk dSPACE software which can barely supervise the correct working of the system. The described test-bed comprises two Smart Meters, both separately used for the two simulated Household, and an insulation transformer interfaces to the grid connection. A precise scheme of the complete architecture is given in Fig. 5.2.

The power electronics system in the setup for one Household is depicted in the upper part of the figure. The converters are powered by a DC power source. An LCL filter is mounted and used to limit the harmonic pollution of the generated waveforms at the AC output terminals. In the setup, out of each pair of power converters, one is employed for the emulation of the non-shiftable part of the load profile, while the other is employed for the emulation of the shiftable part of the load profile. Both converters are then attached downstream the LCL filter and then, to a single Smart Meter that detects the total load consumption of the emulated household. It is important to underline here that the Smart Meter has been installed with its terminals in inverted position. In this way, a power injection into the network is to be regarded

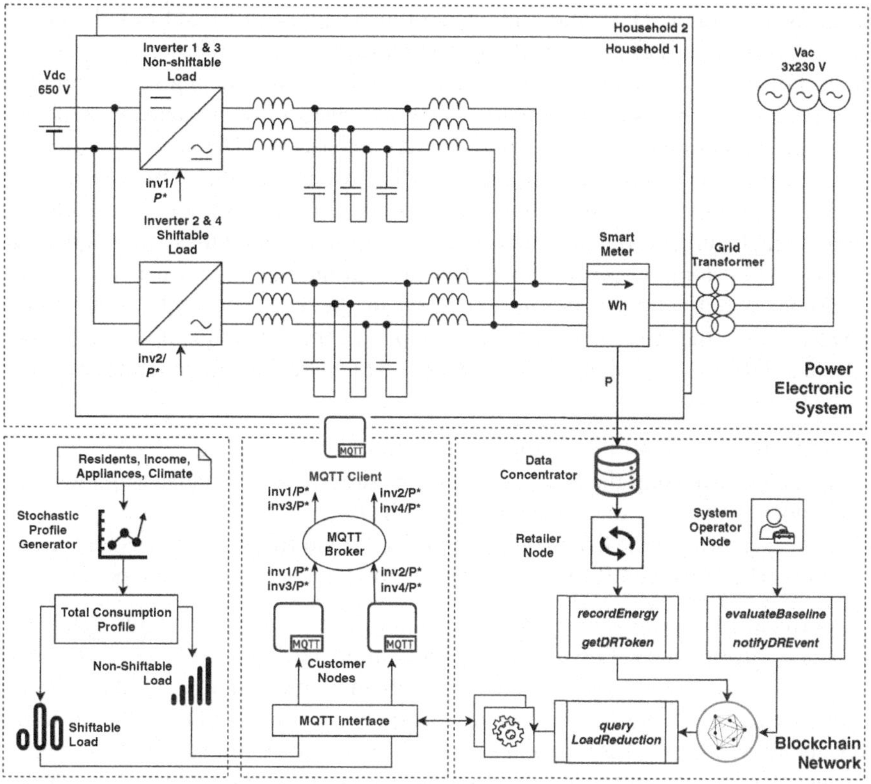

Fig. 5.2 Conceptual diagram of the inverters connection for one customer, blockchain network and SM architecture [1]

as energy load for the emulated households. According to the architecture outlined in Fig. 4.7, a set of four peer blockchain nodes have been deployed following the hardware installations and the Smart Meter to create a blockchain network as shown in the box at the bottom right of Fig. 5.2. As a result, the system operator node and the retailer node run on one computer machine and the two customer nodes run on the setup computer machine. A copy of the Demand Response Smart Contract and a group of customized client Apps were also installed on each node in order to allow every consumer in the network to interact with the blockchain. To keep the test setup as general-purpose as possible, and make easier the integration with third-party commercial solutions, the client Apps were designed as RESTful web services with a stable communication on the back-end with the blockchain network. The general-purpose programming language Node.js was employed to implement these Apps. Figure 5.2 shows the way in which the client Apps and the essential functions can be used to interact with the blockchain. The identities of the retailers are managed by the grid operator. The same grid operator also triggers the baseline calculations (*evaluateBaseline*) and the execution of DR events (*notifyDREvent*). As a result, the grid operator App makes available a RESTful interface for registering retailers identities into the blockchain network. The grid operator also implements a module for triggering periodically the calculations for the baseline. The latter is carried out in the evening. One more RESTful service is then used for posting DR events, in this way, the grid operator publishes the start time, the end time and the requested load curtailment. In turn, the retailer can manage the customers' identities. The retailer also uploads the customers' consumption measures on the blockchain network, which, in the considered setup (*recordEnergy*), reside on the data concentrator. The retailer can also trigger the Smart Contract function to issue the digital tokens for taking part to Demand Response programs (*getDRToken*). Three are the functionalities provided by the retailer App:

- a RESTful service for the customers' registration;
- a module for reading the Smart Meters' measurements and registering them into the blockchain every day, that is run periodically;
- a module to trigger the evaluation of digital Demand Response tokens for the customers when Demand Response events are implemented, that is run periodically.

As a result, the nodes of the customers are simply responsible for acquiring the load reduction information set by the grid operator after starting or triggering a Demand Response event (*queryLoadReduction*) and its implementation using the shiftable loads that are available for the Demand Response event. Therefore, the customer App regularly checks if a Demand Response event has been prompted for a given day and identifies the shiftable part of the load profile accordingly. The customer App is also responsible for triggering th elocal actuation of shiftable loads. This change is carried out by means of the interaction with the MQTT interface which transmits the profiles of consumption to the power converters for the suitable emulation.

Each time a user invokes a Smart Contract function through the client App, a transaction is executed. The transaction information is then shared among the blockchain network peer nodes by means of the "gossip data dissemination protocol" [2]. Peer nodes use this protocol to send data on the communication channel adaptively.

The 'gossip messaging' is continuous and each peer node on a specific channel continuously receives real-time and consistent data from multiple peer nodes. Each 'gossip' message is digitally signed, letting peer nodes sending malicious (i.e.: false) messages to be identified in a simple way. Peer nodes experiencing delays, the partition of the network or other issues causing the loss of blocks can be eventually synchronised with the most recent ledger status by contacting the peer nodes storing the missing blocks. This communication protocol is based on the share of the workload for the operation of transactions. In this way, it is possible to optimise the performance of the blockchain network, considering security and scalability:

- it is possible to manage the identification of peer nodes and their integration in the channel, constantly detecting available and offline peer nodes;

- it is possible to share the data from the ledger among all the peers of a channel letting the data synchronisation solely for those peers that have experienced a loss of blocks;
- it is possible to update the most recently connected peer nodes very quickly on the changes that occurred to the ledger throughout their disconnection.

To create the aforementioned household load profiles a stochastic model for the composition of residential loads was used [3]. Such model provides disaggregated one minute resolution data for a given household comprising appliances, and considering whether inhabitants are residents or vacationers, the income level and the surrounding climate characteristics, which determines the users' habits. The type of appliance included in the household are selected based on the probability of ownership. Among these, in the part which non-shiftable, different appliances can be found: refrigerators, televisions, laptops, pc, stoves, microwaves, oven, iron, coffee maker, toaster, water heater, etc. As a result, only washing machines, dryers and dishwashers are taken as shiftable loads.

5.2 Setting up of the Blockchain Network

Before performing the DR event, it is needed to set the blockchain network. For this applications the network is composed by the following components:

1. PC0, that hosts:
   - A Certification Authority (CA);
   - An Orderer Service Node;
   - A DSO peer (Peer0);
   - A CouchDB instance (couchDB0);
   - An App Client for Peer0;
2. PC1, that hosts:
   - A Utility peer (Peer1);
   - A CouchDB instance (couchDB1);
   - An App Client for Peer1.
3. PC2 (the setup PC), that hosts:
   - A Customer1 peer (Peer2);
   - A CouchDB instance (couchDB2);
   - An App Client for Peer2.
   - A Customer2 peer (Peer3);
   - A CouchDB instance (couchDB3);
   - An App Client for Peer3;

PC0, PC1 and PC2 are personal computers. The first step is to install Hyperledger Fabric on all the computers taking part to the experiment. This very simple procedure is described in the official Hyperledger Fabric documentation and must be repeated on every machine [4]. Once Hyperledger Fabric is installed on all the peer nodes and the network components described above are running, the next step is to generate the communication channel among the peer nodes, where they can all authenticate with their own identities. Subsequently the digital identities of the users that will take part to the network are generated. The developed network is composed of two organizations, part of the "DRconsortium". The identities of the retailer (Org1) and of the end users (Org2) are managed by the MSP of the grid operator. An X.509 certificate links identities to public keys using a digital signature and these certificates are distributed to end users. Listing 5.1 shows a snippet of the NodeJS code used register, create and import the identity of the Customer1 into a folder called *wallet*.

Listing 5.1 Snippet of the code used for user's registration.

```
const secret = await ca.register(
    {
        affiliation: 'org2',
        enrollmentID: 'customer_1',
        role: 'client',
        attrs: [
        {name: "ID", value: "smartmeter:20360162", ecert: true},
        {name: "IDprovider", value: "utility:1234", ecert: true}
                ]
    }, adminIdentity);

const userIdentity = X509WalletMixin.createIdentity('Org2MSP',
    enrollment.certificate, enrollment.key.toBytes());
wallet.import('customer_1', userIdentity);
```

In the first part, in addition to the affiliation, the user's name and role, two attributes are specified: the first identifies the user with the SM ID, while the second identifies the authority that provides the identity to the user. The second part generates the X.509 certificate with the identification data of the user and places it inside the *wallet*. These two attributes are essential to implement the privacy of data shared on the network as explained in the paragraph 4.3. The smart meter ID is used to allow the customer to access only his/her own consumption data or earned tokens, while the utility ID is used to prevent the customer from accessing the data of a customer with the same ID but provided by another retailer. For example, listing 5.2 shows a snippet of the purpose-developed SC that gives permission only to the retailer to use the function.

Listing 5.2 Privacy implementation in the SC through the attribute based access control.

```
// ==== Get the Attribute value for the Retailer ====
ID, ID_found, err := cid.GetAttributeValue(stub, "ID")
    if err != nil {

            return shim.Error(err.Error())
        }
        fmt.Println("ID found : ", ID_found)
        fmt.Println("ID is : ", ID)

// ==== Check if the Attribute matches the value ====
var allowedREs allowedUsers
var jsonResp string
k:=0
allowedREsFromState, err := stub.GetState("allowed REs")
if err != nil {
        jsonResp = "{\"Error\":\"Failed to get state for "+
        ID + "\"}"
        return shim.Error(jsonResp)
        } else {
                err = json.Unmarshal([]byte(allowedREsFromState),
                &allowedMOs)
    for _,value := range allowedREs.IDs {
        if ID == value {
                continue
        } else {
                k+=1
            }
    }
    if k == len(allowedREs.IDs) {
                return shim.Error("You are not allowed to
                record Energy values: only the Retailer can
                call this method")
    }
}

// ==== Check if the Smart Meter is enabled to work on this BC ====
var allowedMeters allowedUsers
h:=0
allowedMetersFromState, err := stub.GetState("allowed Meters "+ID)
if err != nil {
        jsonResp = "{\"Error\":\"Failed to get state for "+args[0]+
        "\"}"
                return shim.Error(jsonResp)
        } else {
                err = json.Unmarshal([]byte(allowedMetersFromState),
                &allowedMeters)
    for _,value := range allowedMeters.IDs {
        if args[0] == value {
                continue
        } else {
                h+=1
        }
        }
        if h == len(allowedMeters.IDs) {
                return shim.Error("This Smart Meter is not enabled
                to work on this BC or the owner is another Market
                Operator")
    }
```

The first part gets the retailer ID attribute, the second checks if the attribute contained in the X.509 certificate of the retailer matches to the one stored on the blockchain by the network manager, in this case the DSO. Finally, the last part verifies whether the SM, so the customer, for whom the retailer intends to store the energy value is enabled to work on the blockchain.

Once all the digital identities of users have been generated and they are enabled by the network manager to work on the blockchain, the SC is installed and can run on each peer and the blockchain network is prepared to operate. In a Hyperledger Fabric network there are two different types of peers depending on whether the SC is installed or not. In the application developed, each peer hosts a copy of the SC, but in wider platforms it may happen that a number of other peer nodes do not host a copy of the SC. A peer node can execute a SC if the same SC is installed on the peer node, otherwise it can access the Smart Contract installed on another peer node by connecting to a communication channel. In this way, all peer nodes can validate and later accept or refuse transactions on their copy of the ledger. However, only peers with a SC truly installed can participate in the transaction approval process, as explained in paragraph 4.3, which is crucial for the generation of valid transactions. In Hyperledger Fabric, peers can take on multiple roles based on how the network is configured or realized. There are two basic types of peers:

1. *Committing peer*: every peer node of a channel is a committing peer. It receives blocks of generated transactions, which are then validated before being appended to the own ledger copy;
2. *Endorsing peer*: each peer of the network can be an endorsing peer if it has an SC installed. Nonetheless, to be an endorsing peer, the SC on the peer node must be used by an App Client to create a digitally signed transaction return.

In this context, a realistic scenario could be one in which the customers are committing peers, so they own only the App Client and a copy of the ledger, but to interact with the ledger they trigger the SC of the endorsing peer owned by the DSO, the retailer or the aggregator. In this way clients do not need computing power to run the SC.

A SC approval policy identifies organizations whose peer should digitally sign a generated transaction before it can be appended on the committing peer's ledger copy. Endorsing peers are divided into two categories:

1. *Leader Peer*: if an organization has more peers in a channel, a leader peer is a node that takes the lead for distributing transactions from the orderer node to the committing peer nodes;
2. *Anchor Peer*: if a peer has the necessity to communicate with a peer in another organization, it can use one of the anchor peers designated in the channel configuration for that organization. An organization may have no anchor peer or it may have more than one anchor peers.

A peer can be a committing peer, an endorsing peer, a leader peer and an anchor peer simultaneously. Only the anchor peer is not strictly needed, but it is convenient to have always a leader peer and at least one endorsing peer and a committing peer. In the considered scenario all the peers are committing, endorsing and leader peers, but only the DSO is an anchor peer because it need to communicates with both organizations.

5.3 Blockchain Workflow for a DR Program Test in a Lab Environment

In the developed application, two main roles are taken into account: the grid operator, which requests the DR event for balancing generation and consumption, and the energy retailer that offers services to the customers. In other scenarios both entities can be unified in the DSO. The DR event is executed through the following steps:

1. *Each day*, the **retailer** transmits the customers' load profiles data to the blockchain by means of the SC, taking them from the DSO data hub. Moreover, the grid operator prompts the SC to compute the baselines and the availability profile of the end users;
2. *The day before the DR event*, the **grid operator** notifies the blockchain network about the day, time window and total load reduction, which is distributed automatically by the SC to the end users following their calculated baselines and their availability profile, according to 4.9;
3. *The day of the DR event*, the **customers** read from the blockchain the desired load reduction computed by the SC in order to meet the grid operator's request. Their actual consumption for that day is also registered as described in point 1;
4. *The day after the DR event*, the **retailer** prompts the SC for customers' prizing with the aforementioned utility tokens. It was assumed that the SC gives them only if customers' consumption during the DR event meets the desired load reduction, ignoring the possible penalties if the consumption does not meet it.

In order to consider the variability of the load profile due to the seasons or the customer's behaviour variation over time, a whole time horizon of 36 d was taken into account for carrying out the experiment. As generating 36 d of consumption profiles using directly the hardware setup involves a time-consuming process which is not necessary, the experiment was divided into three steps:

1. *Pre-allocation*: In the first step, 33 d of consumption profiles were registered into the blockchain with the same resolution of the SM (15 min).
2. *Emulation of everyday behaviour*: In the second step, two days of standard load profiles for the two considered households have been emulated by means of the outlined testbed. To reduce three times the time to carry out the experiment (8 h required for 24 h simulation), the SMs were set to register the five minutes average power. The power set points of the inverters outputted by the model with 1-minute resolution were refreshed with a frequency of 20 s. Moreover, a time settling routine was coded into the retailer App Client so that the 5-minute resolution data reading from the SMs was turned into a 15-minutes resolution data as an effect of the downsizing.
3. *Emulation of Demand Response event*: On the last day emulated, the experimental set-up reproduced a DR event generated by the grid operator. In the emulation, the end users responded by moving the shiftable loads as needed. This operation is carried out by the customer's App Client interacting with the MQTT interface

which transmits the power profiles, so that the profiles of the shiftable appliances are reallocated.

To automatize all the process, a MATLAB script controling the whole process was implemented. This script communicates with the four Apps Client installed at each peer node by using the MATLAB RESTful web services functions 'webread' and 'webwrite'. Moreover, the MQTT library was used to regularly update the set points for the power profiles during steps 2 and 3.

5.4 Test of the Physical Setup for DR Event Execution

Initially, the system must register the load profiles using the Smart Meters, and these latter can be accessed by the retailer node. In the considered scenario, after the 'preallocation' phase lasting 33 d, this method can be used for the two days of normal behaviour (days 34th and 35th) and the day of the Demand Response event (day 36th). A 24 h profile of Household 1 for a normal day is presented in Fig. 5.3.

As it can be seen, in Fig. 5.3a the consumption profile is divided into two components which are sent as power references to a pair of inverters (1 and 2 for Household 1, and 3 and 4 for Household 2). The first inverter (Inv 1) generates the non-shiftable part of the Household consumption profile, while the second inverter (Inv 2) is responsible for emulating the shiftable component of the load. The same for Household 2 and inverters 3 and 4.

Following the power electronic system architecture presented in Fig. 5.2, the output of each pair of inverters is connected to a common SM and later to a common point of coupling with the grid. Therefore, the SM is only capable of measuring aggregated consumption (shiftable and non-shiftable components). Furthermore, while the resolution of the power profiles sent to the inverters is 1 min, the SM takes 15-minute resolution average power samples at the output of the system.

Figure 5.3b shows a comparison between the expected aggregated consumption of the inverters 1 and 2 and the actual power profile recorded by SM 1 for day 34th. These traces have been obtained by the retailer App Client by accessing the DSO data hub and have been subsequently inserted in the blockchain network. It can be observed how the power recorded by the SM totally matches the expected set points, although with a lower resolution.

Concerning the quality of the emulation, the average absolute error between the expected and the recorded profiles for the 24 h test was approximately 10 W. Moreover, the total daily energy consumption was around 1% lower than expected. These errors are probably a result of power losses in the system. Each inverter controls its output power after the LCL filter to compensate for passive losses, yet the use of a common AC/DC source and the parallel connection of the inverters can lead to small circulating currents. These could be avoided by using, for instance, individual isolation transformers, nevertheless, their impact is so small in our scenario that no measure was taken.

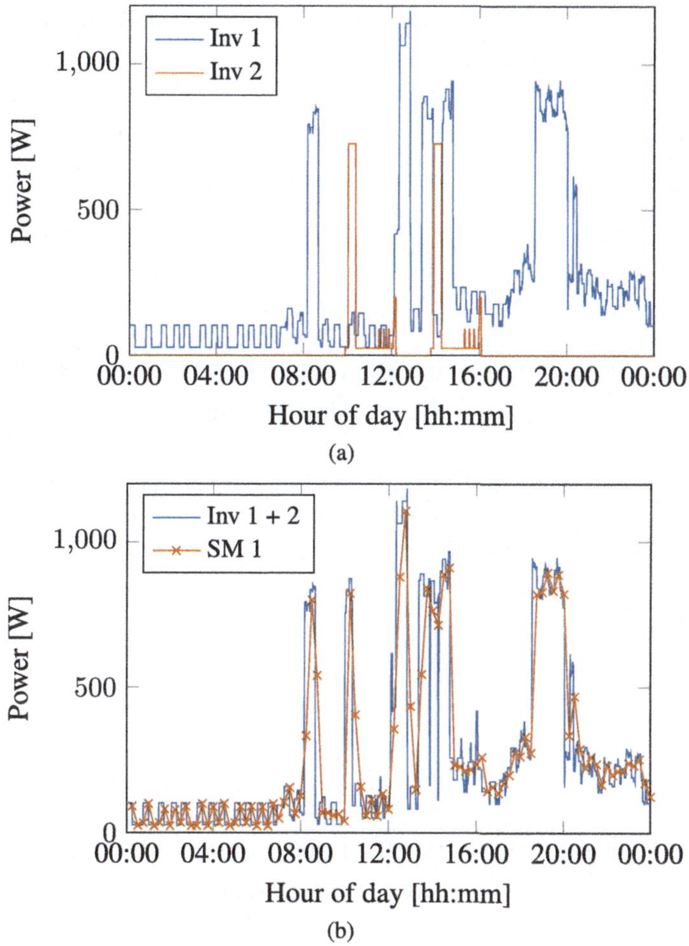

Fig. 5.3 Profiles sent to inverters and recorded by SM into the blockchain, for non-shiftable & shiftable loads (**a**) and total profile (**b**) for Household 1 [1]

5.4.1 Baseline and Optimal Power Reduction Calculations

Following the 35th day, the grid operator prompts a DR event. Therefore, a DR event corresponding to a power reduction of 300 W on average for 15-minutes in the time frame between 07:00 h and 19:00 h is requested by calling a function of the SC designed on purpose. At this point, 35 daily power profiles for each household have been registered on the blockchain and the baseline is computed with these information. Such baselines were computed by the SC using Eqs. 4.2 and 4.3 assuming Y = 35, X = 21 for the weekdays and X = 14 for the weekends baseline. In particular, the day chosen for the DR event is a weekday. Figure 5.4 shows the computed base-

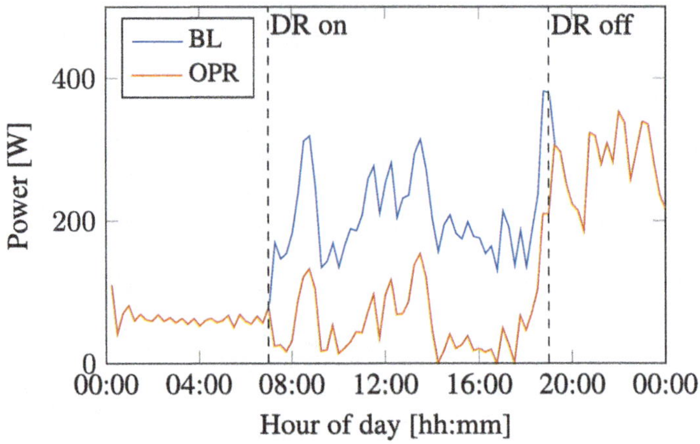

(a)

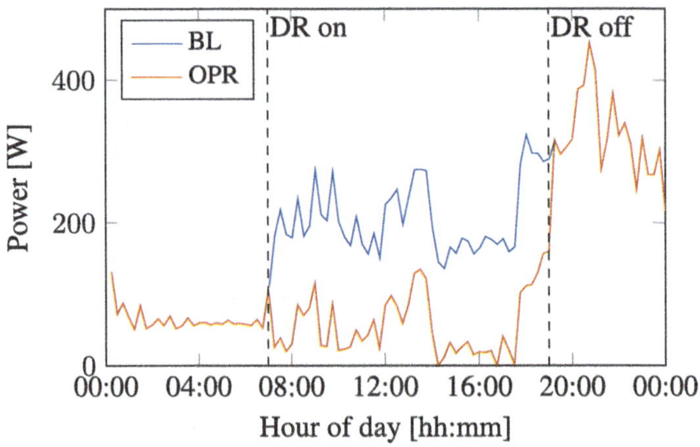

(b)

Fig. 5.4 Baseline calculations for day 35th and OPR for the following day with DR event for Household 1 (**a**) and Household 2 (**b**) [1]

line (BL), such as the optimal power reduction (OPR) obtained for each household. For this test it was assumed that households make available the wole local shiftable power, in this way, their availability profile is equal to 1 and the OPR appreciated by the SC, following Eq. 4.7, is taken as proportional to the baseline of each household.

It can be seen, that baselines differ each other as each household has its own behavioural patterns. As an example, it can be seen that even if the peak of the loads occur around the same hour for both consumers, household 2 has a greater demand peak in the evening as compared to household 1. On the other hand, household 1 has a greater consumption in the central hours of the day.

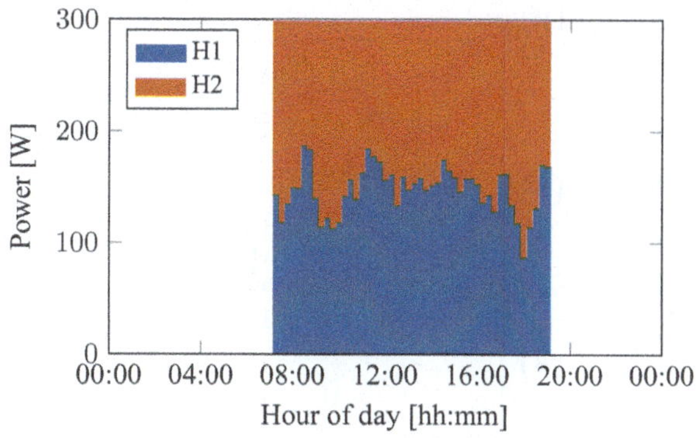

Fig. 5.5 Proportional reduction requested to each household [1]

As a result, the OPR algorithm coded in the SC proportionally shares the required reduction between the two end users. This is shown in Fig. 5.5, in which the 300 W reduction is proportionally shared every 15 min to Household 1 (blue bars) and Household 2 (orange bars).

5.4.2 DR Event Certification and Reward System

The previous section has outlined how the OPR is generated for each household. In this section, the behaviour of the user along the DR event day and its consequences on the remuneration are evaluated. Figures 5.6 and 5.7a present the load profiles of the two households during the DR event day, considering separately non-shiftable and shiftable loads, as well as the aggregated samples taken by the SMs. In order to show meaningful test cases and results, at first only household 1 was considered to meet the DR program.

In Figs. 5.6 and 5.7a, the green line is the shiftable demand. Household 1 has repositioned the shiftable loads to the hours right ahead the DR event prompt to meet the requested power reduction. On the other hand, the shiftable loads in household 2 are also turned on during the DR event.

This is translated to different remunerations for each user as shown in Figs. 5.6 and 5.7b, which represents the 15 min tokens awarded to each users according to Eq. 4.10. Household 1 could mantain its consumption limited for almost all the time in which the DR event was taking place, as a result, as a whole, 188 DRtokens were awarded. Household 2, instead, incurred a high loading for longer time as compared to household 1 and, as a result, the remuneration was only 147 DRtokens.

It should be noted, at this point, that even if a power reduction was requested by the grid operators, both consumers still have high demand and consumption during the DR event. This was anyhow expected as in the considered scenario, the available flexibility can only derive from shiftable loads. This example has evidenced that the grid operator should also take into account this problem, and what are the realistic

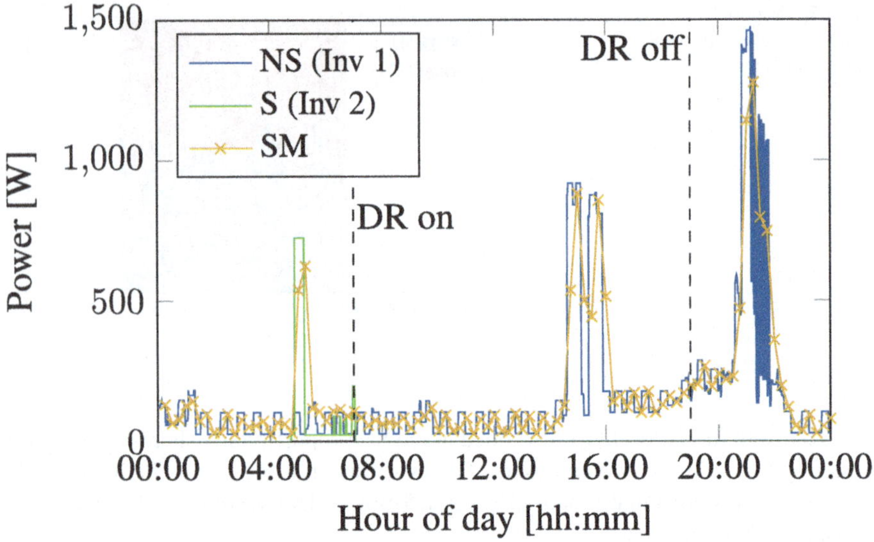

(a) Household 1. Non-shiftable (NS) and shiftable (S) profiles sent to the inverters for the DR day and meterings from the SM.

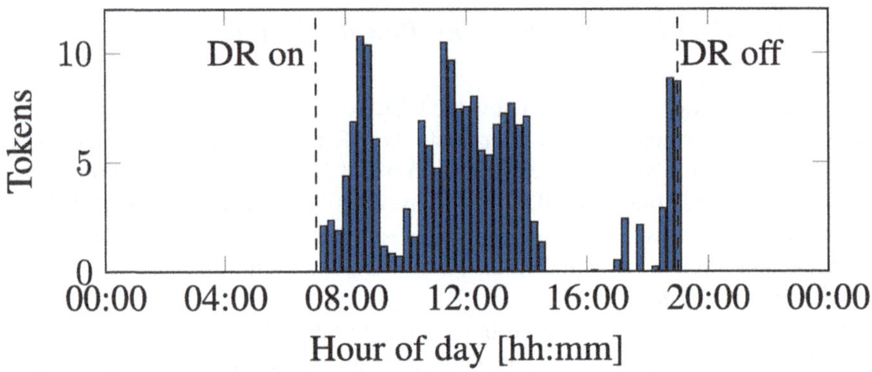

(b) Household 1. Number of tokens awarded for the DR response in 15-minutes intervals.

Fig. 5.6 Response of household 1 to the DR event and rewards given after the day [1]

possibilities to attain a load reduction under certain circumstances. Through the described setup an experimental real blockchain platform was tested at laboratory scale, although end users were simulated by means of inverters controlled by a real-time control platform. The DR signal is in this case transmitted by the DSO (grid operator) on the blockchain and is employed by the control system to change the power set point provided to the inverters to create a load profile that is compliant with the DR event. In a real world application, this task can be implemented by an Energy Management System (EMS) hosting the load management logic and owned

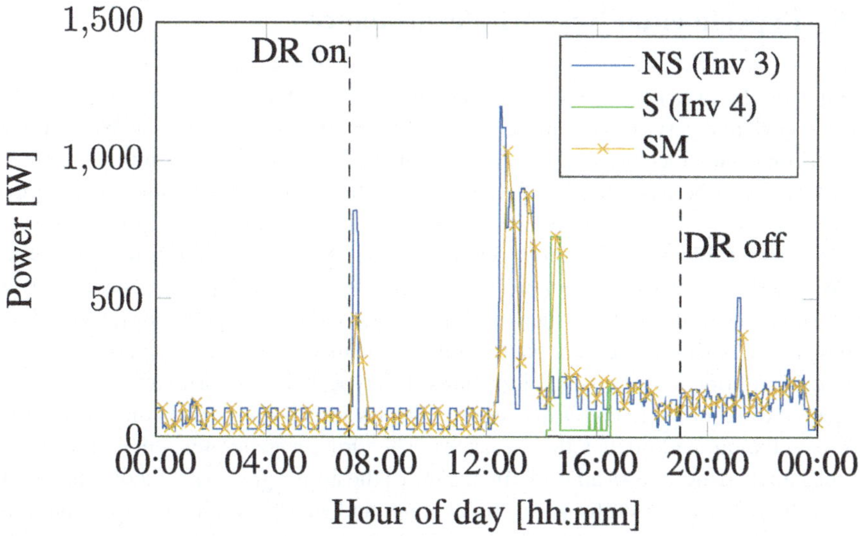

(a) Household 2. Non-shiftable (NS) and shiftable (S) profiles sent to the inverters for the DR day and the records of the SM.

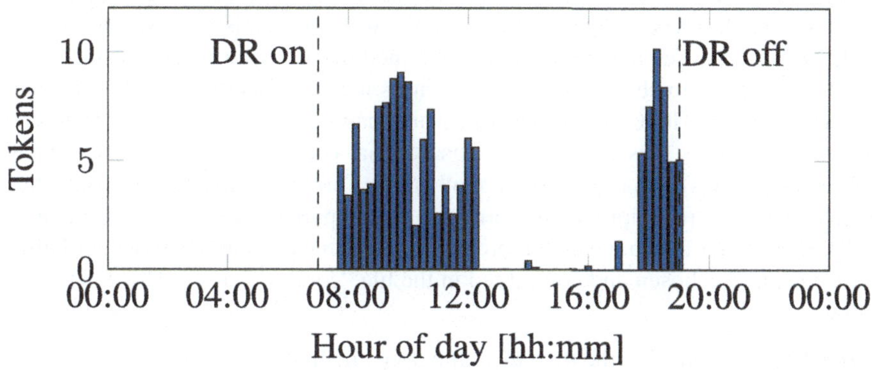

(b) Household 2. Number of tokens awarded for the DR response in 15-minutes intervals

Fig. 5.7 Response of household 2 to the DR event and rewards recognized afterwards [1]

by the customer and external to the blockchain. The load management logic is of course based on the DR signal transmitted by the DSO.

For the laboratory scale experiment here described no EMS was used. However an EMS able to handle local loads at low cost and that can be easily interfaced with the blockchain to meet the requirements of load modulation following a DR event was developed later and is described in the last section.

5.5 Experimental Network Performance

Both in public or permissionless blockchain, such as Bitcoin, anyone can take part to the network and carry out a transaction. In Bitcoin, thanks to a Proof of Work (PoW) based consensus mechanism, the creation time of one block is ten minutes and the block size can be assumed fixed to 1 MB. In this case, the maximum throughput of transaction processing is between 3.3 and 7 transactions per second. The confirmation of 6 blocks takes about one hour time. As an example, Ethereum uses a PoW-based consensus that can process about 25 transactions per second. In private or permissioned blockchain, such as Hyperledger Fabric, the generation time of one single block is shorter, relying on a simpler consensus processes based on faster and less computationally intensive algorithms. In Hyperledger Fabric, transaction messages are transmitted to the Order service. The Orderer then acquires the transactions on that channel and queues the messages. The Orderer generates a new transaction block and distributes it to all peers by means of the above gossip protocol. The gossip protocol lets the peers interact in the channel and sends the channel logs and data in a scalable way. In the public Bitcoin blockchain, all transactions are managed by means of sequential operations, collected in blocks and then added to the ledger. This sequential process does not gain any performance benefit when using more powerful hardware. In Hyperledger Fabric, the consensus mechanism is executed by the Ordering service. The latter is designed in a modular way and is fully connectable. It is then possible to choose scalable consensus mechanisms (Solo, Kafka) for particular use cases. The Ordering service, which can be arranged as a set of Orderer nodes, can process messages. It also ensures that each Ordering process receives transactions and creates blocks in the same order. This event-driven synchronization arrangement ensures improved performance as compared to a public blockchain.

The parameter to change and improve the performance of an Hyperledger Fabric platform can be chosen and configured in the file "configtx.yaml". These are listed below.

- *BatchTimeout*, defines the time to wait before creating a new block;
- *MaxMessageCount*, defines the maximum number of transactions to be inserted in a block during the batch time;
- *AbsoluteMaxBytes*, defines the absolute maximum number of bytes allowed for transactions serialized during a batch time;
- *PreferredMaxBytes*, defines the maximum number of bytes allowed in a batch. A transaction larger than the preferred maximum bytes will result in a batch larger than the preferred max bytes.

To attain higher throughput, it is possible to set *MaxMessageCount* to a higher value. On the other hand, as more computational power is needed, the block increases in size and more bandwidth is required. Generally speaking, by modifying the parameters in configtx.yaml it is possible to increase the transaction throughput, but this may change from case to case based on the application. In this particular application, the authors used the default parameters indicated in Table 5.1. In this way, the

Table 5.1 Used parameters and execution time of SC function for the experimental tests [1]

Blockchain parameter	Value
BatchTimeout	2 s
MaxMessageCount	10
AbsoluteMaxBytes	99 MB
PreferredMaxBytes	512 kB
SC function	Average execution time [s]
recordEnergy	10.2 per load profile (96 energy values)
queryLoadProfile	0.065
evaluateBaseline	12 per customer considering 35 days
queryBaseline	0.065
notifyDRevent	2.3
queryLoadReduction	0.065
getDRToken	2.3
queryEarnedToken	0.065

maximum number of transactions per second is set to 25, namely 50 transactions per batch time. Considering the above discussed scenario, and that the network was implemented using the internal telecommunication network of the Microgrid Laboratoy (with very low latency), the following Table 5.1 shows the time needed for each type of transaction or query carried out by invoking the different functions of the DR smart contract.

Concerning memory consumption, it has been demonstrated that the size in bytes of a transaction for the recording of a power value or DRtoken of a customer is 5 kB, so one block consisting of 10 transactions has a size slightly greater than 50 kB (51.2), while the size of a transaction for recording the baseline, or a DR event, is 10 kB. In general, for the tests executed in this work, the memory consumption can be considered negligible, but considering a real scenario, with hundred of thousands of end users, the size of the chain could become not negligible. In this case, clients with not enough storage capacity could participate to the blockchain as 'light clients', i.e. installing only the App Client, invoking the SC and accessing the ledger on the nearest peer.

5.6 Energy Management System Integration

An Energy Management System, EMS, is a system capable of automatically managing the loads of a user based on weather conditions, energy price signals or user routines in order to optimize energy consumption. The deployment of an EMS connected to the above described blockchain architecture and relevant functionality tests were carried out at the Smart and Microgrids lab and the SnappLab laboratories at the University of Palermo. In this case, virtual machines reproduced the same blockchain architecture described above. The load profiles of customers were recreated and recorded on the blockchain through the use of a Matlab code in order to avoid the use of the experimental setup used in the previously described application. To ensure effective simulated user response to a DR event, the DR signal has been sent by the system operator in the blockchain network and has been thus considered as EMS input data. In this aim, an EMS based on a Non-Intrusive Load Monitoring (NILM) [5] system was implemented. The operation of the EMS and the NILM algorithm is based on historical data. This approach has already been applied in previous literature, such as in [6]. The NILM algorithm was able to disaggregate the total load, recognize the running appliances and store information on the real and reactive power consumption profiles of the appliances and their typical usage profile along the day. The energy measurements have been then processed by the EMS. The latter, in ordinary service, shifts the flexible appliances (indicated by the user) to off-peak hours according to an economic optimization, minimizing the user's costs related to the electricity supply. In the scenario here considered, the integration of the EMS with the blockchain allows to efficiently manage DR events. When the system operator notifies a load reduction request on the blockchain, the SC transmits to the EMS the calculated customer baseline and the extent to which that specific customer must reduce its consumption in the DR event time frame. These data are sent together with the starting time and the duration of the DR event. As described above, the SC shares the whole request considering the availability profile of the customers. The latter represents the customers' ability to meet the request from the grid operator, and evaluates the customers' reward for the service provided. Following the availability profile of the customers, the SC arranges a ranking of their availability. For each DR event, the distribution function of the whole request following the availability profiles is computed. Also in this case, it is considered that taking part to the DR service is ruled by a contract setting an incentive for the customer if he/she can meet the request and, moreover, it is assumed the payment of penalties when the request from the grid operator is not met by the customer. The more compliant the load modulation with the request, the higher the incentive or reward. Under these terms, the customer's participation to the DR program generates an economic benefit, which is here quantified in terms of DRtokens. The objective function of the algorithm coded in the EMS for each generic customer, is shown in Eq. 5.1. The objective function is composed by the cost of the electricity supply, while the incentive for the participation to the DR event, is considered as an inequality constraint, as explained below.

$$min\, C_{el,supply,h} \cdot P_{optim,h} \quad \forall\, h \in T \tag{5.1}$$

In the above equation, $C_{el,supply,h}$ is the supply price of electrical energy following the on-peak, mid-peak and off-peak hours and $P_{optim,h}$ is the actual consumption of the customer during a generic hour h of the day as well as while the i-th DR event is occurring. For ease of notation, even if all quantities of the formulas reported below are related to a single customer, to the generic hour h of the day and to the i-th DR event, the subscripts were omitted. Suppose the customer is purely a consumer (without generation capability), when a DR event occurs, the EMS algorithm gets from the blockchain the OPR evaluated by the SC together with the baseline P_{BL}, in order to evaluate the target load (the reduction of load to perform), as in Eq. 5.2.

$$P_{target} = P_{BL} - OPR \tag{5.2}$$

In this case, the baseline is also acquired from the blockchain. In general terms, although the EMS can also assess historical consumption trends of the user, this may slightly differ from the baseline recorded on the BC, which is shared with all parties.

$$P_{BL} \cong P_{fix} + \sum_{m=1}^{M} P_{flex,m} \tag{5.3}$$

In Eq. 5.3, P_{fix} is the load due to non shiftable appliances and P_{flex} is the load absorbed by each of the M shiftable appliances following the user's habits. The non shiftable and shiftable loads are then identified by the NILM algorithm analyzing the customer consumption habits, and it then identifies which load can be delayed. The optimization algorithm in the EMS schedules the shiftable appliances according to the cost minimization criterion, attaining a so-called "optimized load profile". If the rated power of each shiftable appliance is denoted with $P^*_{flex,m}$, the optimized profile can be formulated as:

$$P_{optim} = P_{fix} + \sum_{m=1}^{M} \delta_m \cdot P^*_{flex,m} \tag{5.4}$$

where δ_m is a vector of Boolean variables (one value for each hour and each appliance) equal to 0 if the appliance is off and equal to 1 when it is on. As already said, taking part to the DR event has been here schematised within the optimization algorithm by means of an inequality constraint rather than as part of the objective function. As shown in Eq. 5.5, it is possible to formulate a the fact that a user can partially respect the DR flexibility request within a predefined tolerance:

$$\Delta P_{DR,h} \leq \varepsilon \tag{5.5}$$

where $\Delta P_{DR,h}$ is the difference between the optimized load profile and the target profile.

$$\Delta P_{DR} = \left| P_{optim} - P_{target} \right| \tag{5.6}$$

Considering Eqs. 5.2, 5.4, 5.5 and 5.6 together, and putting into evidence the term with the Boolean variables, this constraint can be expressed as:

$$\sum_{m=1}^{M} \delta_m \cdot P^*_{flex,m} \leq \varepsilon - P_{fix} + (P_{BL} - \Delta P_{BL}) \tag{5.7}$$

Two more constraints were considered in the optimization algorithm, so as to let the shiftable loads to be moved as compared to a standard schedule avoiding load reduction or peak shaving (Eq. 5.8) and to ensure that the hourly load is lower than the contractual power (Eq. 5.9).

$$\sum_{h=1}^{T} P^*_{flex,m} \cdot \delta_{m,h} = \sum_{h=1}^{T} P_{flex,m,h} \;\; \forall \; m \in M \tag{5.8}$$

$$P_{perf,h} \leq P_{contract} \;\; \forall \; h \in T \tag{5.9}$$

The result of the EMS algorithm is the recalculated optimal combination of shiftable loads along the day as a DR event occurs, while meeting the reduction request and minimizing the cost beared by the customer. This is mathematically formulated by the vector of δ_m. As a result, the customer can avoid to pay a penalty for not meeting the request and the entire process is carried out automatically. In order to show the possible interaction between the BC and the EMS, a case study with a reduced number of appliances was developed. A randomly created domestic load profile was used for the P_{fix}, while, for the shiftable load, a dishwasher and a kettle were considered. Rated power of shiftable appliances were taken from [7]. Load profiles for the user historical consumption (data recorded on the EMS and analyzed through the NILM algorithm) are shown in Fig. 5.8, with a resolution of 15 min. More parameters used in the case study are presented in Table 5.1. The electricity supply prices are derived from the economic conditions set by the Italian energy authority for the first three months of 2020 [8] (Table 5.2).

After the optimization, the shiftable appliances were rescheduled as shown in Fig. 5.9. As it can be observed, the kettle turning on was delayed of a given number of hours, so as to reach the mid-peak period and to set aside money. As far as the dishwasher operation is concerned, instead, as this appliance is usually in operation during the DR event, its powering was delayed during the off-peak hours to meet the system operator request, and to earn money as well.

In the following paragraphs, the interaction between an energy blockchain platform and a domestic EMS, minimizing the cost for the customer related to the energy supply, is described. The cooperation of these systems allows the customer to automatically meet the DR request from the system operator and to have this service offered to the grid operator certified. This feature is very important, as the involvement of residential customers in DR programs is affected from a high risk of non compliance, causing many uncertainties. Nonetheless, even if the described EMS system uses a NILM algorithm to identify the running appliances, the automated shifting of the loads would require the installation of smart plugs as actuators on each flexible appliance. It is worth to note that, in the case study above presented,

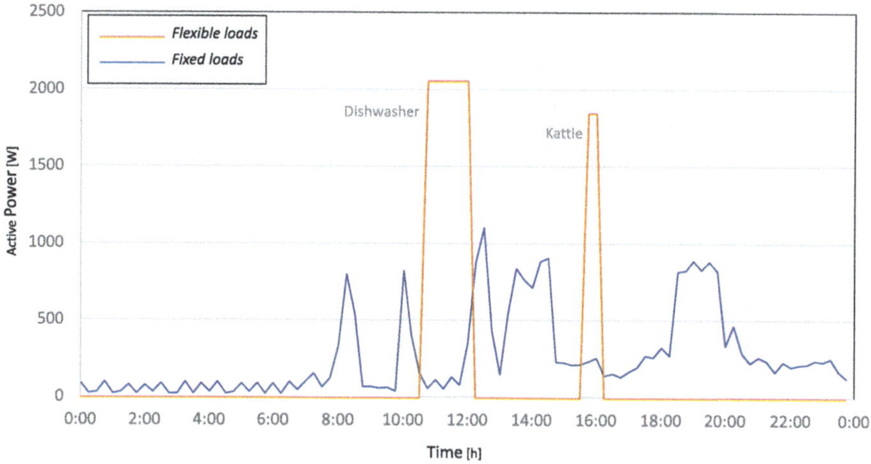

Fig. 5.8 Historical customer trends of fixed and flexible loads

Table 5.2 Parameters employed for the case study

Parameter	Symbol	Value
Number of flexible appliances	M	2
Dishwasher rated power	$P^*_{flex,1}$	2060 W
Kettle rated power	$P^*_{flex,2}$	1853 W
User contractual power	$P_{contract}$	3300 W
Reduction order	OPR	1000 W
Start time of DR event	t_{start}	10:30 A.M.
Duration of DR event	t_{DR}	2.25 h
On-peak electricity price	$C_{el,supply,on-peak}$	7.666 c€/kWh
Mid-peak electricity price	$C_{el,supply,mid-peak}$	6.799 c€/kWh
Off-peak electricity price	$C_{el,supply,off-peak}$	6.799 c€/kWh

meeting the DR request would mean the shifting of the dishwasher to off peak hours, whereas the delayed powering of the kettle to mid-peak hour was implemented only for the economic reasons and a benefit for the customer. By means of this action, the maximum daily peak of the user grew from 2.4 to 2.7 kW, even though it occurs in a different time of the day. Moreover, the actual load reduction for the customer was much lower than the amount requested by the DR program, since the shiftable load rated power was about double the requested reduction, causing a load valley instead of a peak. These two issues suggest that the DR management is the result of a trade-off between the user's and grid operator's needs, thus including further uncertainty. To consider this, the role of the blockchain could become even more important if, after each customer's EMS algorithm has found the local optimum setting of flexible loads, all customer programs are sent back to the blockchain and to the SC to verify

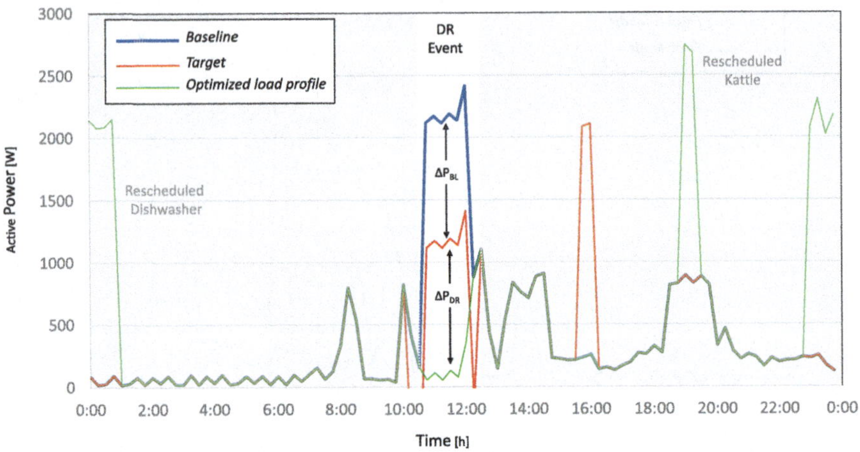

Fig. 5.9 Rescheduling of flexible appliances after optimization

whether the cumulative load reduction is in line with the system operator's needs. This would give BC the further role of a peers manager, in addition to being a virtual aggregator.

5.7 Integration in the MSD

DR programs are important tools supporting the continuous correct operation of the power system due to RES penetration, as it allows to adapt the energy demand to the supply in a short time. During peak load periods and in response to market signals, users can make self-produced energy available or reduce their consumption against remuneration or other forms of financial incentives. Today in Italy, DR services allow commercial and industrial users to help with electricity regulation by participating in the Dispatch Services Market, MSD. As already introduced in Chap. 2, with Resolution 300/2017/R/EEL of the ARERA, the Italian TSO Terna opened the MSD for the first time to small power plants with non-programmable production, loads and storage systems (Virual Units of Mixed Aggregation, UVAM). Currently, UVAMs represent the benchmark form of aggregation in Italy. UVAMs allow users to make available flexibility by modulating consumption or production in a way that is led by DR programs. Managed by Balance Service Providers, BSPs, UVAMs offer capacity to the MSD and are remunerated on the basis of the variation of the exchanged energy derived on the MSD (€/MWh) plus a fixed amount (in €/MW/year) proportional to the offered capacity, and awarded by auction. According to Terna's UVAM MSD Regulation [9], UVAMs can provide the following dispatching services:

a. resolution of congestion, increasing and/or decreasing;
b. rotating tertiary reserve, increasing and/or decreasing;
c. tertiary replacement reserve, increasing and/or decreasing;
d. balancing, increasing and/or decreasing.

UVAMs provide the services indicated above if they are able to modulate loads with the following features:

- within **15 min** of receipt of Terna's dispatching order for the services referred to in points (a), (b), (d);
- within **120 min** from the receipt of Terna's dispatching order for the services referred to in point (c);

and support such modulation for a period at least equal to:

- 120 min for the services referred to in letters (a), (b), (d);
- 480 min for the services referred to in letter (c).

In addition, the BSP interface device with Terna systems must be able to send every 4 s the total power input/output at the input/output points included in the UVAM and receive the balancing orders through the file transfer mode provided by the IEC 60870-5-104 protocol, thus using a mainly Ethernet transport network [10].

The blockchain network deployed and described above allows the aggregation of users in a transparent way and without the need for a BSP, generating UVAM able to provide flexibility to the power network. The experimental test was conducted to demonstrate the possibility of certification and direct execution of DR events between system operator and end users, with the aim of generating a new system that involves more domestic users to the MSD. In fact, unlike UVAM, it was considered that the DR signal is notified in the network, on the day before the event, but the results obtained are encouraging for applications on current UVAM. The execution times of the main SC functions shown in Table 5.1 (2.3 s for a write operation on the blockchain and 0.065 s in reading) are compliant with the timescale envisaged by Terna. With such timing, UVAMs can provide balancing services on the MSD. Even better if the DR signal is used by an EMS capable of automating user response. Regarding the IEC 60870-5-10 protocol imposed by Terna for data exchange, the blockchain may become compliant by integrating the support of such a protocol into the blockchain clients. In such a case, Remote Terminal Units used in current UVAMs interface Terna's control systems [11] could also integrate such blockchain clients. This last aspect needs further studies and insights that will be developed in future work.

References

1. G. Sciumè, E.J. Palacios-García, P. Gallo, E.R. Sanseverino, J.C. Vasquez, J.M. Guerrero, Demand response service certification and customer baseline evaluation using blockchain technology. IEEE Access **8**, 139313–139331 (2020). https://doi.org/10.1109/ACCESS.2020.3012781
2. Hyperledger, Gossip data dissemination protocol, in *Hyperledger-Fabricdocs Documentation—Release Master* (2017). Accessed 29 Oct 2020. [Online]. Available https://www.hyperledger.org/wp-content/uploads/2017/08/Hyperledger%7B%5C_%7DArch%7B%5C_%7DWG%7B%5C_%7DPaper%7B%5C_%7D1%7B%5C_%7DConsensus.pdf
3. E. Palacios Garciá, *Stochastic Modelling and Integration of Distributed Energy Resources in the Smart Grid*. Ph.D. dissertation (Dept. of Electronic and Computer Engineering, University of Cordoba, Cordoba, Spain, 2018). [Online]. Available https://helvia.uco.es/handle/10396/16204
4. Hyperledger, *Hyperledger-Fabricdocs Documentation, rRelease Master: Install Samples, Binaries and Docker Images*. Accessed 3 Nov 2020. [Online]. Available https://buildmedia.readthedocs.org/media/pdf/hyperledger-fabric/latest/hyperledger-fabric.pdf
5. G.W. Hart, Non intrusive appliance load monitoring. Proc. IEEE **80**(12), 1870–1891 (1992). ISSN: 1558-2256. https://doi.org/10.1109/5.192069
6. R. Godina, E.M.G. Rodrigues, E. Pouresmaeil, J.C.O. Matias, J.P.S. Catalão, Model predictive control home energy management and optimization strategy with demand response. Appl. Sci. **8**(3) (2018). https://doi.org/10.3390/app8030408
7. C. Beckel, W. Kleiminger, R. Cicchetti, T. Staake, S. Santini, The eco data set and the performance of non-intrusive load monitoring algorithms, in *Proceedings of the 1st ACM Conference on Embedded Systems for Energy-Efficient Buildings, ser. BuildSys '14* (Association for Computing Machinery, Memphis, Tennessee, 2014), pp. 80–89. ISBN: 9781450331449. [Online]. Available https://doi.org/10.1145/2674061.2674064
8. ARERA, *Condizioni economiche per i clienti del mercato tutelato* (2019). Accessed 7 Nov 2020. [Online]. Available https://www.arera.it/it/dati/condec.htm
9. Terna, *Regolamento recante le modalità per la creazione, qualificazione e gestione di unità virtuali abilitate miste (uvam) al mercato dei servizi di dispacciamento*. Accessed 16 Nov 2020. [Online]. Available https://download.terna.it/terna/Regolamento_UVAM_MSD_8d803bd91884d36.pdf
10. Terna Spa, *Requisiti tecnici minimi delle apparecchiature per la rilevazione e comunicazione delle misure e per la gestione degli ordini di bilanciamento*. Accessed 16 Nov 2020. [Online]. Available https://download.terna.it/terna/Allegato%202_8d803be56f0fd2d.pdf
11. Terna, *Criteri di connessione al sistema di controllo terna* (2018). Accessed 16 Nov 2020. [Online]. Available https://download.terna.it/terna/A.13_8d7b5fc76904df7.PDF

Chapter 6
Blorin Blockchain Platform

G. Sciumè, E. Riva Sanseverino, P. Gallo, A. Augello, G. Sciabica, and M. Tornatore

Abstract Based on what explained in the preceding sections, this chapter reports an overview of the Blorin project, implementing Demand Response and Vehicle To Grid services supported by a permissioned blockchain platform. We describe how the use of the Hyperledger Fabric blockchain for developing our blockchain-based energy management platform. Subsequently, the configuration of a distributed network is considered, supported by a multi-host environment in which all actors of the Blorin project have a role and give a specific contribution. Hyperledger Fabric (HLF) is a blockchain that permits participants accessing and managing their transactions in the platform through suitable smart contracts, which can be used also for validating data. The special feature of this project is that it allows in a modular way the access to different services supported by specific applications. Thus the energy services taken into account are only two of the possible ones. Participants will thus take part in the platform through plug-and-play services and an approach that allows, with scalable solutions, the participation of a large number of end users, while preserving privacy.

6.1 Blorin Architecture

The substantial difference with other blockchain systems consists in the way in which all members subscribe to the platform, namely through a trusted Membership Service Provider (MSP). As a result, the validation of the transactions and the safety of the network do not need "proof of work" algorithms and this implies a substantial difference as compared to open systems, so called permissionless, that allow unknown identities to access the platform with writing capabilities. Among the various novelties, this project offers different mechanisms and tools that can be combined allowing an easy integration with existing technologies. For this reason, the ledger data can be stored in multiple formats and the consensus mechanisms can be substituted by different MSPs. HLF is used in many other application sectors

G. Sciumè (✉) · E. Riva Sanseverino · P. Gallo · A. Augello · G. Sciabica · M. Tornatore
University of Palermo, Viale delle Scienze, Building 9, 90128 Palermo, Italy
e-mail: giuseppe.sciume01@unipa.it

G. Sciumè et al. (eds.), *A Practical Guide to Trading and Tracing for the Energy Blockchain*, UNIPA Springer Series, https://doi.org/10.1007/978-3-030-96607-2_6

and, thanks to its modular architecture, it is considered the first platform supporting smart contracts written in generic programming languages such as Java, Go and Node.js, rather than the specific programming languages from Ethereum like Solidity. Another feature of HLF is that the participants to the platform are exposed to each other being not anonymous. The network works even in environments that are not trusted, thus we consider a Zero Trust approach. Unlike other platforms, HLF supports modular consensus protocols that can be plugged and unplugged allowing a more flexible customization. Therefore, differently from consensus protocols that are prized by mining on a public blockchain, it is possible to exploit consensus protocols that do not require an underlying cryptocurrency, otherwise used for the smart contracts execution. Avoiding the need of a cryptocurrency, for letting the platform work, reduces the risks and the attacks and, in absence of crypto-mining operations, it is possible to operate the platform at reduced energy consumption as compared to other systems.

The current development of the Blorin network includes 7 nodes, each of which represents specific interests in the energy scenario (Fig. 6.1).

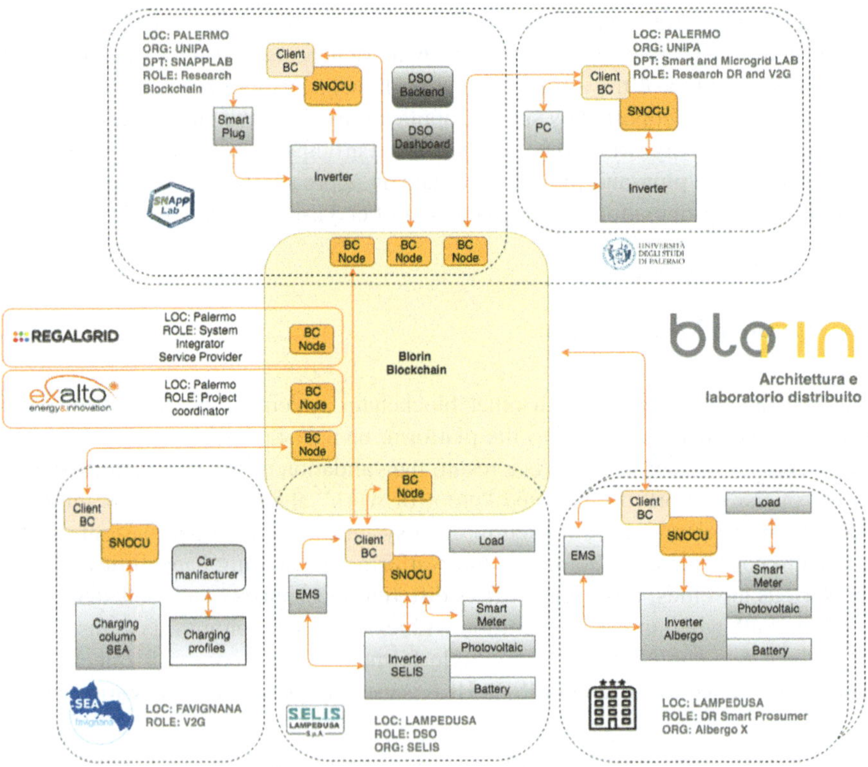

Fig. 6.1 Blorin architecture and main actors

As it can be seen from 4.7 the facility at the University of Palermo comprises 2 nodes, at the SNAPPLab (Security Network Applications and Positioning Laboratory) and at the SMG (Smart and MicroGrid Lab). Any of the involved elements maintains a node, as follows. The SNAPPLAB aims at developing and controlling the entire communication infrastructure among the different peer nodes and the blockchain itself. The Smart and Microgrid LAB provides hardware components to test the energy services provided by end users (including V2G) and their tracing using a suitable metering system. Two more nodes are maintained by Regalgrid, a system integrator and service provider, and Exalto, the coordinator of the Blorin project. Other key involved actors are the grid operators. The distribution company SEA in the Favignana island takes care of managing the V2G service and the distribution company SELIS takes care of managing the DR service in the Lampedusa island. Moreover, in Lampedusa, several prosumers and consumers are going to be involved. At the University of Palermo and in Lampedusa we, respectively, simulate/emulate and test new scenarios including Peer-to-Peer transactions and Energy Communities. Prosumers will interact at the same level and EVs recharge infrastructures will be able to interact with consumers, having the chance to manage improved services. In this way, it will be possible to improve Quality of Service (QoS) and Quality of Experience (QoE) for end users in order to maximise their benefits. Users save money and time and earn while technically supporting the smart grid; the grid operators obtain improved ways to manage the electrical service and make their generators work at the highest efficiency possible. The blockchain technology allows to manage energy exchanges among prosumers, to aggregate them and deliver balancing services.

6.2 Blorin Research Platform

The Blorin blockchain network has been deployed using Docker as virtualization platform. Docker was created with the aim of making autonomous the distribution of applications under the form of portable, light and self sufficient containers, with the possibility to be executed through the cloud or locally [1]. The distribution process of an application is thus reduced at the creation of a Docker image and, through the latter, it was possible to create a container that executes the application inside the Docker engine.

Containers are self-sufficient as they already include all the dependencies from the network, they do not need as a result to execute requests in the host configurations. The latter can indeed be read by any other Docker server, guaranteeing full portability. Moreover the limited memory footprint of containers allows to exploit the services offered by the host Operative System and thus, the hosting machine can run a large number of containers. The use of a blockchain for managing the energy distribution network has some advantages in comparison to the legacy methods, the main one is that to provide more efficient balancing services for the daily load with a possible direct actuation of the management policies in the network.

Using a blockchain allows a greater efficiency in the use of renewables energy sources and a better energy management. The key feature of this project is that it has an optimal replication potential, such technology is indeed not so invasive, but allows to exploit existing infrastructures transparently delivering information to all stakeholders. The new scenarios that will be set up will support the wide implementation of smart energy communities with higher Quality of Service (QoS) and Quality of Experience (QoE), in this way, it will be possible to maximize the benefit for the end users that will save time and money and for the grid operators who will better manage the electrical service. The considered technical solutions have to account for a number of issues:

1. Analysis of regulatory barriers that currently impede the energy exchanges between peers and prosumers and the possible evolution of the legislation at national and international levels.
2. Analysis of possible business models that can be adopted for the energy blockchain, underlining possible risks and advantages for the involved actors.
3. Definition of the management model for the P2P market actors involved:
    - BloRin platform manager
    - prosumers
    - distributors
    - consumers
    - recharge unit owner.

For the development of the Blorin platform several devices have been designed, implemented, and deployed. Smart meters allow the reading of the power flows. We use a couple di IoT gateways: the SNOCU produced by Regalgrid and the EMS prototyped by the University of Palermo. These devices read data from smart meters, write them on the blockchain and run local intelligence (in addition to the distributed ones obtained with smart contracts. Additionally, they apply control actions to actuators according to messages written on the blockchain.

6.3 Blorin at SNAPPlab

The infrastructure at the SNAPPlab laboratory includes two servers, named ‘Levante’ and ‘Ponente’. These are all managed by a Proxmox VE virtualization environment, which is open source and work with virtual machines and LXC containers. Such platform manages Virtual Machines and manages clusters through a web interface, it is useful for the development of a multi-node network allowing the development of the system outlined in Fig. 6.2.

The SNAPPlab laboratory includes not only the Levante and Ponente servers and three workstations with useful nodes for the development of the platform, but also devices such as snocu and Blorin Energy Management System (Blorin EMS). In Fig. 6.3 there are reported four cases, two for Demand Response (DR) and two for

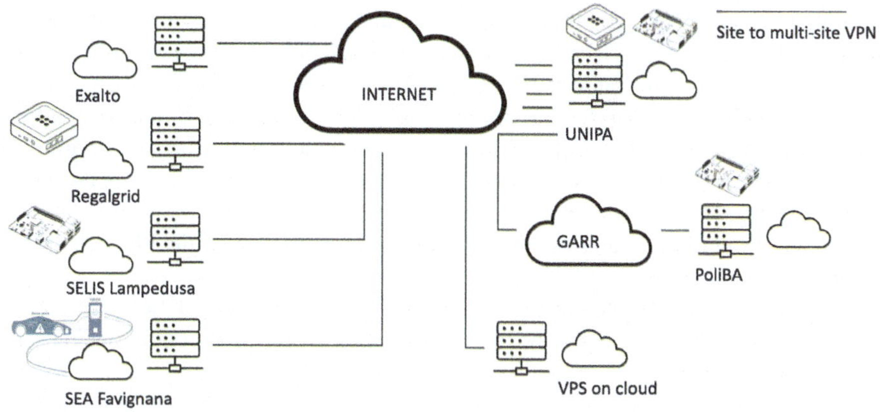

Fig. 6.2 Blorin network infrastructure

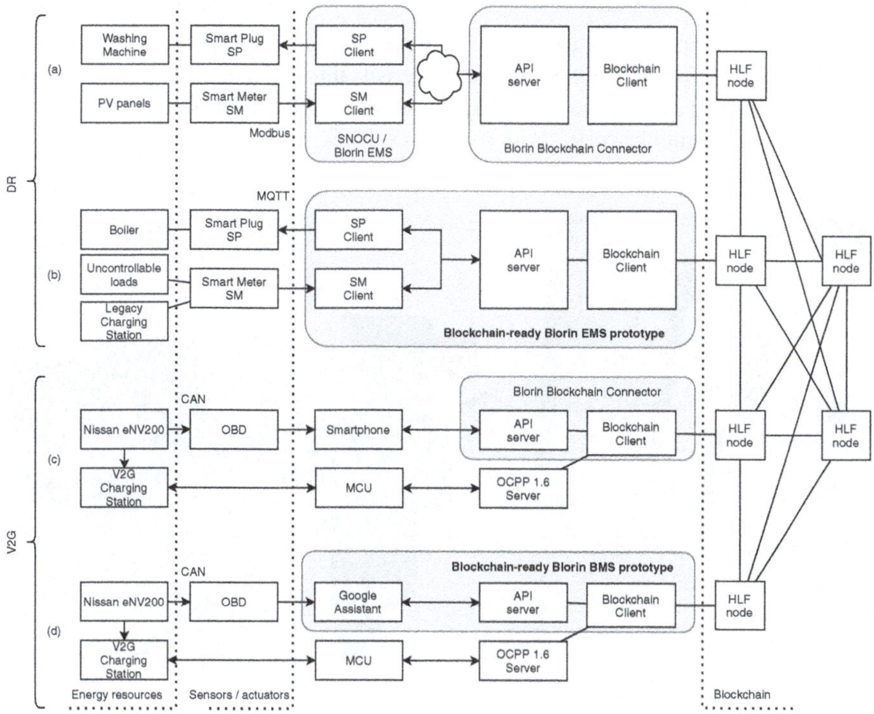

Fig. 6.3 Device perspective and prototypes for DR and V2G, with Blorin EMS and BMS

Vehicle to Grid (V2G). This figure shows the involved energy resources, the monitoring and controlling devices, as well as the involved protocols. In particular, unlike case (a), in (b) it is introduced the Blorin Blockchain-ready Energy management System, which is a prototype that not only saves the installation of an extra device but also permits to apply local policies to the distributed logic for participating to the DR programs. Analogous elements appear in cases (c) and (d). Is composed of a Raspberry Pi, which hosts the client application for monitoring, acquiring data via smart meters and implementing it by executing commands from the blockchain via smart switch.

For implementation, the laboratory is equipped with Meross, Sonoff and Shelly smart devices, which allow for the monitoring of environmental conditions via sensors and in the eventuality of autonomously deactivating the loads.

The logic part is developed on the Blockchain nodes that store all the monitored parameters and in the eventuality control the actuators. Figure 6.4 shows the workstations, the rack that contains the network elements and the servers, the Blorin EMS and one actuator.

These client applications interact with the blockchain through a REST API. The blockchain works on the digital plane and interacts through actuation with the physical one, describing a complete cyber-physical system. SNAPPlab is the VPN endpoint for the whole Blorin network.

Fig. 6.4 SNAPPlab laboratory

6.3.1 *Blorin Network*

6.3.2 *VPN Management*

In terms of security, the VPN allows to solve the problems related to data security that can hardly be guaranteed in public data networks, moreover, it allows to set a private network creating safe tunnels among the LANs. VPNs may help to guarantee authenticity, privacy, reliability and integrity of data transmission. Blorin uses a VPN managed by ZeroTier, a virtualization platform for VPN as a service allowing to connect devices with end-to-end and automatic cryptography. The installation procedure of ZeroTier One and the start of the VPN are simple and can be deployed as follows [2]:

- Creation and verification of an account on https://my.zerotier.com/
- Creation of a network on https://my.zerotier.com/network
- Download ZeroTier on the devices to be added to the network
- Adding devices as members (Fig. 6.5).

Depending on the operating system of the device to be added to the VPN, the procedure is different:

On Windows there is the possibility to download and install the executable ZeroTier One from the official website.

On Linux/Unix/MacOS it is possible to execute the following commands from the terminal:

```
$ curl -s https://install.zerotier.com | sudo bash
$ sudo zerotier-cli join $<ID_NETWORK>$
```

As for Android, it is possible to download the application from the PlayStore: (https://play.google.com/store/apps/details?id=com.zerotier.one&hl=it&gl=US).

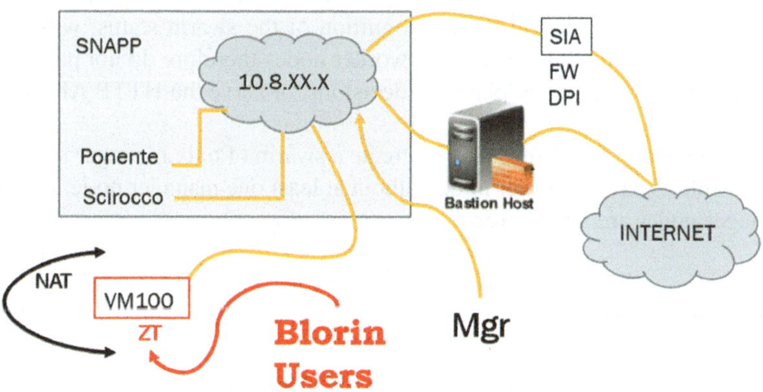

Fig. 6.5 Blorin architecture for VPN management

6.3.3 Docker and Swarm

The Blorin blockchain nodes are deployed through Docker Containers from specific images. Containers can host software that can be packaged and run and the technology to do this is called Docker and software is used to generate the HLF network basic components: Peer, Orderer and Certification Authorities. HLF components can be run as command line executables or as Docker containers. HLF provides Docker images for all its components and a Docker image is a template that is used to instantiate a container just like a Java class is a template used to instantiate an object. A container goes through several states during its lifecycle. First, it is created from a Docker image. Once it is created, the container exists but is not running. The second stage is when container is started, then it is in the running status. While running, the container can be suspended, entering in a paused state, or resumed. Finally, starting from running it can be stopped (with stop or kill commands), then removed.

Containers are self-sufficient because they contain all the dependencies of the network thus guaranteeing full portability. In the network deployment, using a Docker swarm permits managing different containers and planning of loads among containers instantiated in several hosts.

A group of Docker hosts executed in the 'Swarm' mode creates a 'cluster swarm' made of two types of hosts: managers and workers, with different roles that can get the different hosts in the grid. Managers can be considered as the 'guardians' of the state and together they can be considered as a 'consensus cluster' that keeps the recognition of the status of the services in execution of the swarm.

Manager nodes handle cluster management tasks by allowing cluster state maintenance, applying scheduling services, and serving HTTP API endpoints in swarm mode. To take advantage of the fault tolerance features of swarm mode, Docker recommends having multiple handlers so that you can recover from failure of a manager node without downtime. The task is that to guarantee that the state of the swarm really corresponds to that of the developer.

The 'worker' nodes receive and simply execute the activities of the manager nodes and thus do not provide any recognition of the swarm status, with the sole purpose of running the containers. The worker nodes therefore do not participate in the distributed state, make no planning decisions, or serve the HTTP API in swarm mode.

Based on the above, it is possible to create a swarm of only manager nodes, but it is not possible to have a worker node without at least one manager node. By default, all manager nodes are also workers.

6.3.3.1 Quick Start

To initialize the swarm, the following command in the first host is executed, resulting the first network manager [3].

```
$ docker swarm init --advertise-addr=<HOST_IP_1>

$ docker swarm join-token manager
```

With the command join-token manager a token is generated. It allows to add new managers to the swarm, therefore we prosecute with the execution of the following command in the VM that want to join the HLF network.

```
$ <output da join-token manager> --advertise-addr <
    HOST_IP_N>
```

Added all the participants, the first overlay network is created on the first host, allowing the network components to share the same container.

```
$ docker network create --attachable --driver overlay
    blorin-network
```

6.3.4 Hyperledger Fabric in Realistic Contexts

Once executed the above steps, the Docker Swarm is ready to host a multi-host network. The swarm is thus configured and the network can be started generating a Fabric-CA, namely a certification authority (CA) for Hyperledger Fabric that can be used to deliver the functionalities as:

- identities registration
- subscription certificates release
- renewal and revocation of the certificate.

Moreover, Hyperledger Fabric CA is constituted by a server component and by a client and all communications with the server CA of Hyperledger Fabric occur through API REST. Therefore the following commands can be executed to start the network and create one or more channels based on the requested use:

```
$ ./network.sh up -ca

$ sudo ./network.sh createChannel
```

Once the channel is created a copy of the cryptographic material is generated with a consequent start of the docker on all the VMs that must be added to the network, thus exploiting the following commands:

```
$ scp -r <host_name>@<HOST_IP_1>:/home/<host_name>/
    blorin-net/
  blorin-network/organizations/peerOrganizations ./
      organizations

$ scp -r <host_name>@<HOST_IP_1>:/home/<host_name>/
    blorin-net/
  blorin-network/organizations/ordererOrganizations
      ./organizations

$ scp -r <host_name>@<HOST_IP_1>:/home/<host_name>/
    blorin-net/
  blorin-network/system-genesis-block ./

$ ./network.sh up2
```

The joinChannel command allows to join all the peers to the blockchain network.

```
$ sudo ./network.sh joinChannel
```

Once united all the peers to the channel, the network is ready for the implementation of a generic chaincode.

```
$ sudo ./network.sh deployCC
```

6.3.5 *Blorin ClientApp*

To allow the communication of the different nodes with the blockchain network, an API Server was developed in javascript with the functionality to query and write inside the Ledger. This allows to read the data coming from the smart meters through the SNOCU or the EMS devices. The latter, exploiting the commands GET and POST, query the blockchain server and receive control actions considering the implementation logic of the Smart Contracts. After the first start of the blockchain network, it is suggested to execute the subscription of an admin and a user. To do so, the CAs generate the identities given to administrators. It is fundamental to underline here the importance of the identity of the admin that issues the registration. The admin has the ability to send a registration request to the CA server in order to retrieve the registration certificate (CSR). Consequently, once this is done, the admin will have the right to register a new user (the user) by sending a network administrator registration call to the CA server. From the first operations, the private key, the public key and the X.509 certificate are generated. The admin is involved in the process of providing a registration ID and key to the end user. The user can then can log into the CA Fabric client using these credentials. This process both preserves the integrity of the CA because only CA administrators can register users and assign roles and private keys (Figs. 6.6, 6.7 and 6.8).

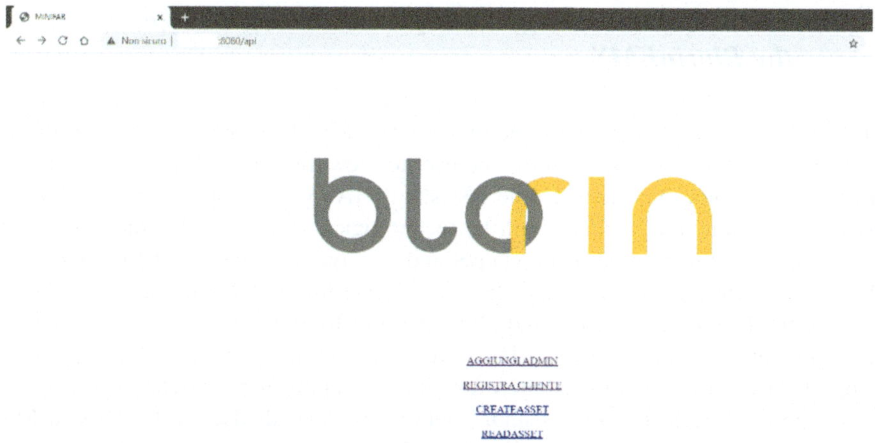

Fig. 6.6 Web APIServer interface

Fig. 6.7 Registration of a new user

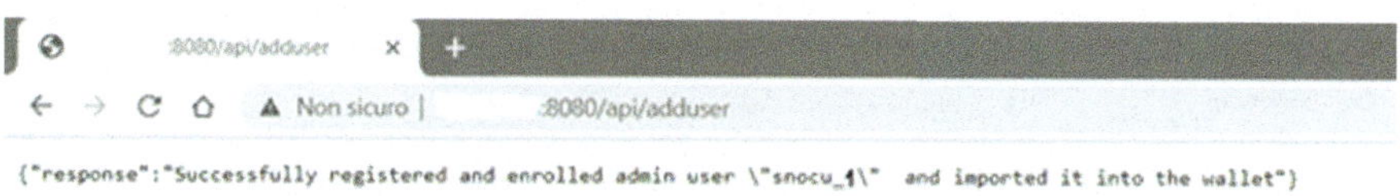

Fig. 6.8 Example of response from APIServer

6.3.6 Control of the Energy Load for the User Through the BlorinEMS

One of the main aspects to combine the control of loads through the decentralized network and at the same time allow the user to view the status of the devices in the house is the Home Assistant. Home Assistant (HASS) is an open-source software based on Python able to control all the technological devices inside the home and at the same time to communicate via get and post requests with the Blockchain via API request, this allows to manage the loads according to the requests of the DSO moving the load independently and allow the user to monitor what is happening in his home and possibly refuse the DR request. Being an open-source product, it is supported by a large community, therefore, it is constantly and constantly updated. To ensure portability, the Home Assistant Container was installed inside the BlorinEMS Raspberry Pi. This allowed to keep the Raspbian operating system with all its features and at the same time take advantage of the tools for home management with HASSIO (Fig. 6.9).

In developing the Dashboard, the user, as in figure, displays the status of smart devices and consumption on the network. The installation of Hass with Docker is done through the following commands:

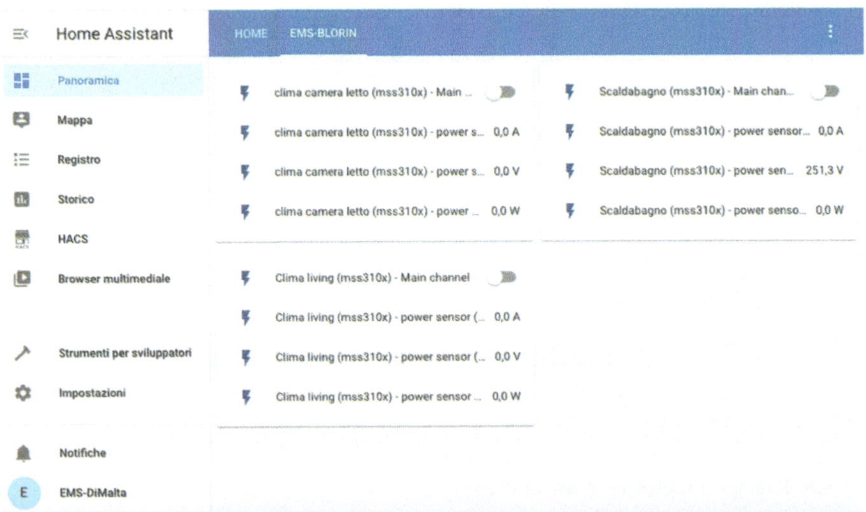

Fig. 6.9 Dashboard viewed by the user

```
docker run -d \
--name homeassistant \
--privileged \
--restart=unless-stopped \
-e TZ=MY_TIME_ZONE \
-v /PATH_TO_YOUR_CONFIG:/config \
--network=host \
ghcr.io/home-assistant/raspberrypi4-homeassistant:stable
```

Once Home Assistant Container is running, Home Assistant can be accessed using the link `http://<host>:8123`. The front-end dashboard, called Lovelace, offers several tabs for viewing information and control devices. The interface is fully customizable by editing the YAML configuration file. Home Assistant acts as a central hub by combining different devices and services in one place and integrating them as an entity. The system through trigger events allows you to create custom routines, to control the energy consumption of the home and to manage various types of alarms.

6.4 Blorin Infrastructure at SMGlab

The SMGLab hosts the prototype of a Smart Prosumer, realized with the objective of emulating a user, both consumer and producer while hosting a blockchain node, to optimally manage production and consumption and to dialogue with the blockchain for the provision of energy services. The selected scenarios are simulated and tested at the SMGlab before being applied to real scenarios deployed in the islands of Lampedusa and Favignana. For this reason, the need to simulate the prosumers that will be part of the new blockchain network arises. The prototype developed for the laboratory emulation of a smart prosumer interacting with one or more blockchain nodes consists of the following components:

1. **switchable resistors**: for simulating step-variable passive loads;
2. **controllable electric engine**: for simulation of active loads;
3. **energy production system**: for simulation of production from a photovoltaic system;
4. **storage system**: to simulate the exchange of energy with the network;
5. **energy management system**: for the coordinated control of all components;
6. **intelligent monitoring and metering system**: for monitoring, measuring load and generation profiles, and communications with the blockchain;
7. **protections and isolation transformer**: for protection against electric shock and over-current.

The consumed or generated energy is monitored through energy meters. The energy production system aims to simulate the presence of a RES production system, in this case photovoltaic, which exchanges energy with the storage system and with the loads, represented in turn by the switchable resistors (passive loads) and the electric motor (active load). The correct management of the inverter, storage and loads allows to obtain the load or generation profiles typical of a prosumer belonging to a microgrid. These profiles are then recorded by the meters that through Modbus RS485 interface communicate with the SNOCU. The SNOCU introduces smartness to the micro-grid prototype, allowing the control of the storage system and interfacing the meters to the blockchain. Thanks to SNOCUs, it is possible to read the consumption or generation data measured by the meters through a platform made available by Regalgrid, and then send this data to the blockchain platform. In this way, it is possible to implement the interaction between prosumer and blockchain, record load/generation data or implement consumption and generation control logics based on energy business models implemented through the blockchain. In the Blorin project, two main business models are considered, respectively for Demand-Response (DR) and Vehicle-to-Grid (V2G). In both cases, the technical objectives required by the network operator are network stabilization and balancing. The two business models are implemented by the blockchain through SCs shared by the blockchain users and are usually executed to implement transaction logic and verify its validity, or to automatically execute transactions if certain conditions are met. In the case of DR, for example, the SC must allow you to: –read data from the Blockchain; –calculate the typical baseline of each prosumer; –publish a DR event; –record and certify consumption/generation during DR events; –calculate the remuneration of the prosumers that have responded to the request.

The SC implementing the DR program therefore aims to modulate loads to limit peak power at certain times of the day. Therefore, the grid operator, having a load forecast available, for each substation makes the desired load/storage modulation schedule and sends the request to the blockchain directed to the loads/storage downstream of the affected substation. Users read the request and act accordingly. Similarly, for V2G, the grid operator sends the request on the blockchain and the V2G charging stations in the area acquire the request and perform the required modulation to balance the upstream energy flow. The developed smart prosumer prototype thus allows testing of both the effectiveness of SCs and the consumption and/or production control system.

6.4.1 Hardware Components

The above cited prototype reproduces a smart grid, independent from the electrical system, installed at the Smart and Microgrids LAB at Ground Floor of Building 9 of the Engineering Department of Palermo University. The system is powered through an IEC 32A 400V socket from one of the existing socket switchboards. The following figure shows the component diagram of the prototype realized to simulate the scenarios described in the previous paragraph (Fig. 6.10).

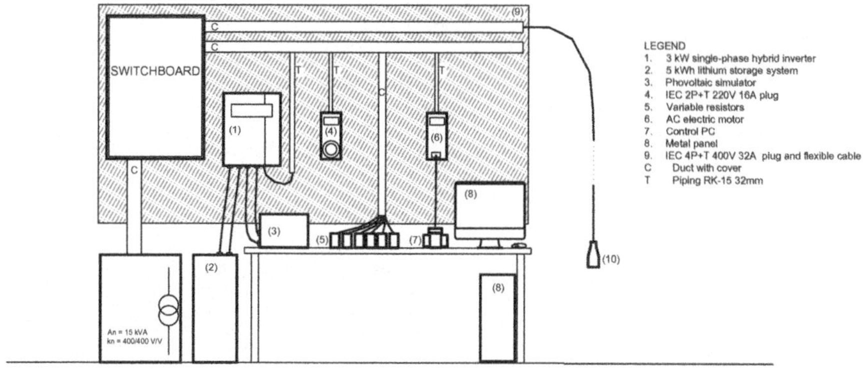

Fig. 6.10 Smart prosumer prototype components diagram

Fig. 6.11 Switchboard for power supply, management and monitoring smart prosumer

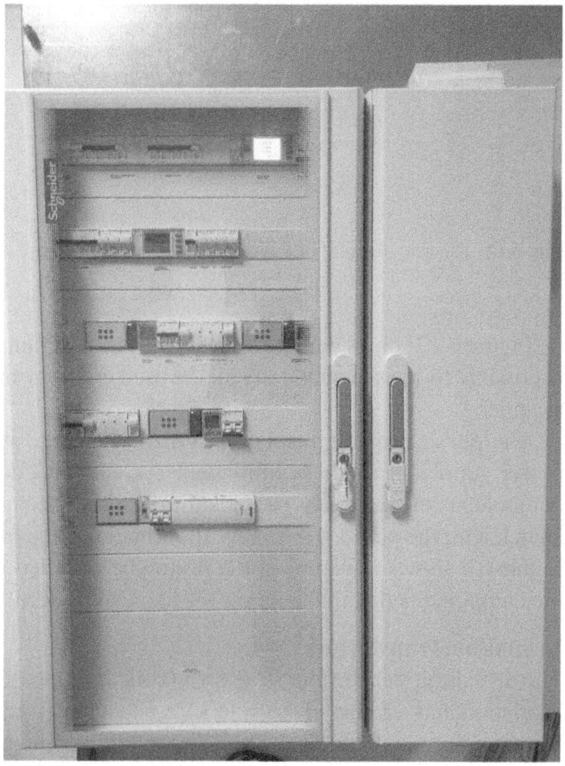

All components are powered through the switchboards shown in Fig. 6.11. This switchboard connects the prototype to the electrical network and contains all the protections, monitoring and control systems (SNOCU).

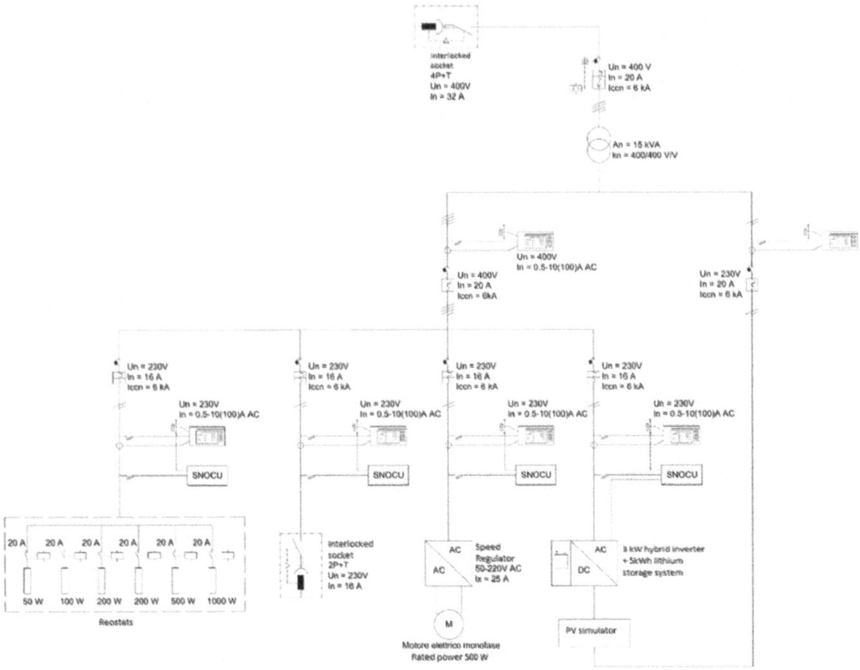

Fig. 6.12 Electric diagram of the smart prosumer prototype

Figure 6.12 shows the single line electrical diagram of the prototype from which it is possible to see schematically all the used components and how they are connected together.

The purpose of the insulation transformer is to isolate the prototype from the power supply in order to limit fault currents and dangerous overvoltages on the components of the prototype. In addition, a release button was included on the main switch with the purpose of timely shutting down the power supply in case of danger. Figure 6.4 shows a picture of the prototype installed inside the lab, from which all the components that make up the system can be seen.

Insulation Transformer

The insulation transformer (Fig. 6.13a) is a three-phase transformer with epoxy resin insulation, has a rating of An = 15 kVA and a transformation ratio kn = 400/400 V/V. The transformer features a Dy11 assembly. As already described in the previous paragraphs, this transformer has the purpose of isolating the power supply circuit from that of the smart prosumer by realizing an IT system (supply system with isolated neutral and grounded utilization, according to Italian technical CEI norms). The choice of the IT system is directly linked to service continuity requirements but is also motivated by users' or operators' safety. The IT system realized presents some peculiarities such as:

(a) (b)

Fig. 6.13 Isolation transformer (**a**) and temperature control unit (**b**)

- neutral isolated from ground;
- non-obligatory disconnection at the first fault;
- signaling at the first insulation fault;
- compulsory trip at the second fault.

The transformer is also equipped with a control unit to control the temperature of the windings as shown in Fig. 6.13b.

Photovoltaic Power Generator

To simulate the presence of a PV generator, an inverter powered by a programmable DC generator is used (Fig. 6.14a). The generator can emulate the typical behavior of a PV system and is sized so that its output parameters respect the coupling conditions with the inverter. The programmable DC generator can be used as a constant voltage source with current limiting or as a constant current source with voltage limiting. The integrated control board allows the variation of voltage values in the range 0–330V, and current values in the range 0–22A, as depicted in Fig. 6.14b. The maximum AC power consumption is 3.1 kW.

The device also has an AUTO-RANGE function that determines automatic switching between two current values. To simulate a photovoltaic curve some reference parameters specified by the manufacturers of photovoltaic panels are required and usually refer to standard conditions (temperature of 25 °C and an irradiation of 1000 W/m^2). As a result, the first step in generating the photovoltaic curve is to choose the PV module to be simulated and enter its specific data into the simulator. The simulation interface also supports a table mode with which you can simulate a custom curve or simulate an existing photovoltaic curve of several panels in series where some panels are shaded. All the necessary parameters for the photovoltaic simulation can be configured using the desktop application through which it is possible to visualize the calculated photovoltaic curve with the entered values and the voltage and current measurements of the power supply.

The inverter used is a HYD 3000 ES hybrid from ZCS, which is combined with a 5 kWh lithium storage system (Fig. 6.15).

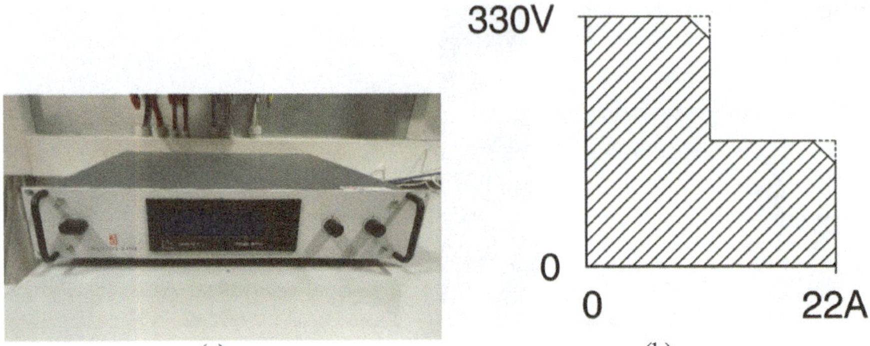

(a) (b)

Fig. 6.14 Photovoltaic simulator (**a**), DC generator capability (**b**)

(a) (b)

Fig. 6.15 ZCS Hybrid Inverter HYD 3000 ES (**a**), 5 kWh storage system (**b**)

The inverter allows the management of the battery charge/discharge mode according to user-customizable programs that can be set remotely, providing good flexibility for both domestic and non-domestic applications. The inverter is designed to work in the presence of the grid (On Grid mode) but also to power privileged loads in the absence of the latter (EPS mode: Emergency Power Supply). Table 6.1 shows the main characteristics of the inverter.

Table 6.1 Inverter technical data

DC input	
Max DC power	3500 W
Max voltage input	600 V
Rated voltage inputt	360 V
MPPT number/strings for MPPT	2/1
MPPT voltage range	90 V/580 V
Max MPPT current	12 A
AC output	
Rated power	3000 W
Max current	13.7 A
Rated voltage	230 V
Rated frequency	50 Hz/60 Hz
Power factor	1
Battery connection	
Rated voltage	48 V
Max charging/discharging power	3000 W
Max charging/discharging current	65 A (programmable)
Temperature range	−10 °C/50 °C
Deep of discharge DoD	0/90% (programmable)

Loads

The passive loads are simulated with variable rheostats of the linear type with slider of ItalOhm, consisting of cylindrical porcelain tubes on which is made a resistive aluminum winding. Moving the slider along the appropriate guide, you can adjust the resistance value of the rheostats and, consequently, the power absorbed by them, simulating different load entities. The choice of these components was made in response to the need for simplicity and immediacy in the simulation of a domestic load as well as the need for strong and long-life components that can withstand the stress of laboratory tests. In addition, the component represents well the behavior of most loads connected to small and medium-sized LV systems that have power factor between 0.95 and 1. There are a total of six 200 Ω rheostats each capable of consuming a maximum power of 1200 W with a maximum current of 2.4 A. The rheostats are connected to the test bench by means of test lead connectors and connected/disconnected by controlling the closing/opening of contactors installed in the switchboard (Fig. 6.16a).

Finally, a single-phase asynchronous motor is also used to simulate ohmic-inductive loads, Fig. 6.16b. The size of the load is regulated by acting on a special speed controller. The size of the motor was chosen to be 0.5 kW, which is compatible with that of the loads present in common domestic users.

The following Fig. 6.17 shows the prototype assembled in the SMGLab.

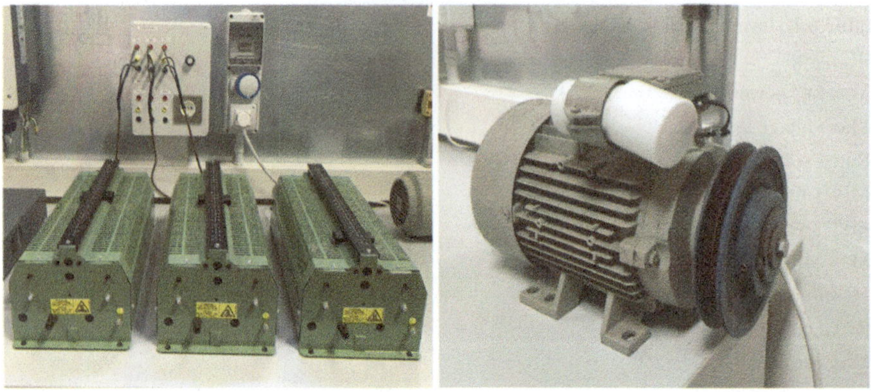

(a) (b)

Fig. 6.16 ItalOhm 1, 2 kW reostats (**a**) and single-phase electric (**b**)

Fig. 6.17 SMGLab smart prosumer prototype

```
'EASTRON_SDM230': {
    'name': 'Eastron SDM230',
    'data_points': [
        { 'modbus_addr': 0x00, 'name': 'L1_V', 'description': "L1 Line (V)", 'type': 'float' },
        { 'modbus_addr': 0x06, 'name': 'L1_A', 'description': "L1 Current (A)", 'type': 'float' },
        { 'modbus_addr': 0x0C, 'name': 'L1_W', 'description': "L1 Power (W)", 'type': 'float' },
        { 'modbus_addr': 0x12, 'name': 'L1_VA', 'description': "L1 Volt Amps (VA)", 'type': 'float' },
        { 'modbus_addr': 0x34, 'name': 'Tot_W', 'description': "Total System Power (W)", 'type': 'float' },
        { 'modbus_addr': 0x38, 'name': 'Tot_VA', 'description': "Total System Volt Amps (VA)", 'type': 'float' },
        { 'modbus_addr': 0x0156, 'name': "Tot_KWh", 'description': "Total Energy (kWh)", 'type': 'float' },
        { 'modbus_addr': 0x46, 'name': 'LineFrequency_Hz', 'description': "Line frequency (Hz)", 'type': 'float' },
        { 'modbus_addr': 0x48, 'name': 'EnergyImported_kWh', 'description': "Energy Imported Accumulated (kWh)", 'type': 'float' },
        { 'modbus_addr': 0x4A, 'name': 'EnergyExported_kWh', 'description': "Energy Exported Accumulated (kWh)", 'type': 'float' },
        { 'modbus_addr': 0x54, 'name': 'N_MAX_A', 'description': "Neutral Maximum Current (A)", 'type': 'float' },
        { 'modbus_addr': 0x0108, 'name': 'L1_MAX_A', 'description': "L1 Maximum Current (A)", 'type': 'float' },
```

Fig. 6.18 Data read from the smart meter with indications from the MODBUS addresses where the data are stored

```
url = 'http://[illegible]:8080/api/createBlorinAsset/snocu_1'
myjson = {"id": "testsdm230", "value": valore}

x = requests.post(url, json = myjson)
```

Fig. 6.19 Extracted from the blockchain client

6.5 Blorin Sensing

6.5.1 Smart Meters and Modbus

The client-meter installed on Raspberry-Pi allows to acquire the data deriving from the smart meter through serial port RS485 following the communication protocol Modbus. Through an adapter USB-RS485 it is possible to read the data from the smart meter "Eastorn SDM230" that already is equipped with the following reading schema of Modbus addresses (Fig. 6.18):

The data, after a subsequent reading in real time of the smart meter, are written on the blockchain exploiting the Client-BC suitably read for data reception in json format (Fig. 6.19).

6.5.2 Reading Data from Blockchain

Furthermore, the data residing on the blockchain network can be read to allow to execute control actions directly on the SNOCU device, the latter occurs querying the blockchain through the Client-BC. The reading of the data coming from the blockchain is possible thanks to the query of single data or directly from the historical data, as described in Figs. 6.20 and 6.21.

{"response":"{\"ID\":\"sdm230-snapp\",\"Value\":\"Thu, 13 May 2021 13:03:58 CEST{\\\"meterType\\\": \\\"Eastron
SDM230\\\", \\\"L1_V\\\": 243.41868591308594, \\\"L1_A\\\": 0.11325851827859879, \\\"L1_W\\\": 16.915019989013672,
\\\"L1_cos_phi\\\": 0.8083510398864746, \\\"L1_VA\\\": 20.925342559814453, \\\"Tot_W\\\": 0.0, \\\"Tot_VA\\\": 0.0,
\\\"Tot_kWh\\\": 29.23200035095215, \\\"LineFrequency_Hz\\\": 50.0, \\\"EnergyImported_kWh\\\": 29.23200035095215,
\\\"EnergyExported_kWh\\\": 0.0, \\\"N_MAX_A\\\": 16.79652976989746, \\\"L1_MAX_A\\\": 0.45441731810569763}\"}"}

Fig. 6.20 Exemplary reading of a single data

Raw Parsed

{"response":"[{\"ID\":\"sdm230-snapp\",\"Value\":\"Thu, 13 May 2021 13:03:58 CEST{\\\"meterType\\\": \\\"Eastron
SDM230\\\", \\\"L1_V\\\": 243.41868591308594, \\\"L1_A\\\": 0.11325851827859879, \\\"L1_W\\\": 16.915019989013672,
\\\"L1_cos_phi\\\": 0.8083510398864746, \\\"L1_VA\\\": 20.925342559814453, \\\"Tot_W\\\": 0.0, \\\"Tot_VA\\\": 0.0,
\\\"Tot_kWh\\\": 29.23200035095215, \\\"LineFrequency_Hz\\\": 50.0, \\\"EnergyImported_kWh\\\": 29.23200035095215,
\\\"EnergyExported_kWh\\\": 0.0, \\\"N_MAX_A\\\": 16.79652976989746, \\\"L1_MAX_A\\\": 0.45441731810569763}\"},
{\"ID\":\"sdm230-snapp\",\"Value\":\"Thu, 13 May 2021 13:03:51 CEST{\\\"meterType\\\": \\\"Eastron SDM230\\\",
\\\"L1_V\\\": 244.712890625, \\\"L1_A\\\": 0.11594069749116898, \\\"L1_W\\\": 17.350683212280273, \\\"L1_cos_phi\\\":
0.8131666779518127, \\\"L1_VA\\\": 21.337181091308594, \\\"Tot_W\\\": 0.0, \\\"Tot_VA\\\": 0.0, \\\"Tot_kWh\\\":
29.23200035095215, \\\"LineFrequency_Hz\\\": 50.0, \\\"EnergyImported_kWh\\\": 29.23200035095215,
\\\"EnergyExported_kWh\\\": 0.0, \\\"N_MAX_A\\\": 16.79652976989746, \\\"L1_MAX_A\\\": 0.45441731810569763}\"},
{\"ID\":\"sdm230-snapp\",\"Value\":\"Thu, 13 May 2021 13:03:44 CEST{\\\"meterType\\\": \\\"Eastron SDM230\\\",
\\\"L1_V\\\": 244.01907348632812, \\\"L1_A\\\": 0.114244744181633, \\\"L1_W\\\": 16.367673873901367,
\\\"L1_cos_phi\\\": 0.7944236397743225, \\\"L1_VA\\\": 20.603208541870117, \\\"Tot_W\\\": 0.0, \\\"Tot_VA\\\": 0.0,
\\\"Tot_kWh\\\": 29.23200035095215, \\\"LineFrequency_Hz\\\": 50.0, \\\"EnergyImported_kWh\\\": 29.23200035095215,
\\\"EnergyExported_kWh\\\": 0.0, \\\"N_MAX_A\\\": 16.79652976989746, \\\"L1_MAX_A\\\": 0.45441731810569763}\"},
{\"ID\":\"sdm230-snapp\",\"Value\":\"Thu, 13 May 2021 13:03:38 CEST{\\\"meterType\\\": \\\"Eastron SDM230\\\",
\\\"L1_V\\\": 244.67752075195312, \\\"L1_A\\\": 0.11420992016792297, \\\"L1_W\\\": 16.42983627319336,
\\\"L1_cos_phi\\\": 0.7954934239387512, \\\"L1_VA\\\": 20.653644561767578, \\\"Tot_W\\\": 0.0, \\\"Tot_VA\\\": 0.0,
\\\"Tot_kWh\\\": 29.23200035095215, \\\"LineFrequency_Hz\\\": 50.0, \\\"EnergyImported_kWh\\\": 29.23200035095215,
\\\"EnergyExported_kWh\\\": 0.0, \\\"N_MAX_A\\\": 16.79652976989746, \\\"L1_MAX_A\\\": 0.45441731810569763}\"},
{\"ID\":\"sdm230-snapp\",\"Value\":\"Thu, 13 May 2021 13:03:29 CEST{\\\"meterType\\\": \\\"Eastron SDM230\\\",
\\\"L1_V\\\": 244.0533447265625, \\\"L1_A\\\": 0.11392262578010559, \\\"L1_W\\\": 16.418773651123047,

Fig. 6.21 Exemplary reading of historical transaction data from the blockchain

In this way, it is possible to query the blockchain and receive useful commands for the execution of control actions coming from the network itself.

6.5.3 *Electric Vehicle*

In the aim of executing a communication between an electric vehicle and the blockchain for tracing the delivery of energy services, a connection using a smartphone has been created (Fig. 6.22). The smartphone collects data from the vehicle through an OBD-II device and uses the Leaf Spy Pro application for sending the data coming from the vehicle to the server so as to exploit the communication between client and blockchain (Figs. 6.23 and 6.24).

Leaf Spy Pro can receive data under the following two conditions:

- Leaf is in 'Ready' mode
- One or more of the three charge lights of the dashboard are on/flashing (during recharge or for about 5 min after having set the recharge timer)

The 'Ready' mode can be activated in the following ways:

- Foot on the brake and turn the ON button (key for start the EV is inserted)
- Foot on the brake and press two times the ON button

To turn off the indicator lights over the dashboard the following steps must be taken:

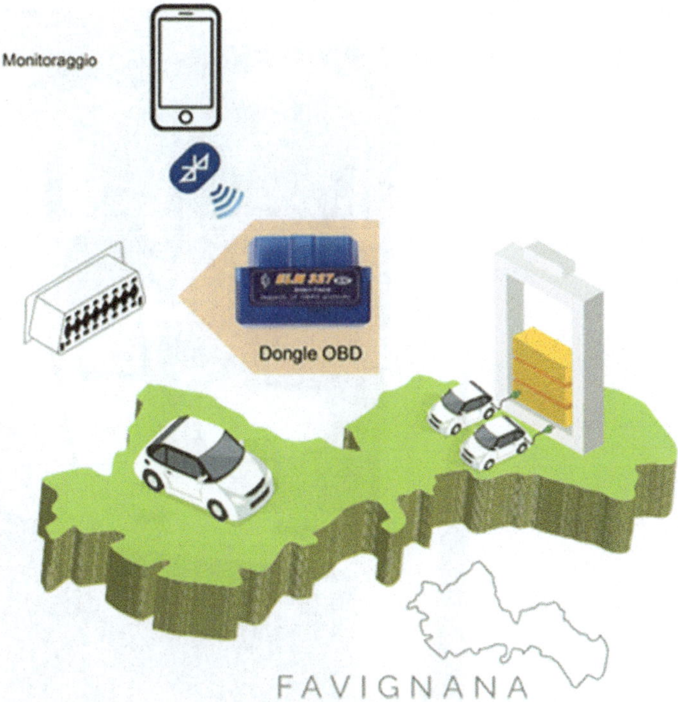

Fig. 6.22 Hardware for data acquisition to the blockchain from an electric vehicle

- Press the override button of the timer
- Disconnect and reconnect the recharge cable
- Start of the timer or activation of remote charge.

Through the OBD-Bluetooth (ELM327) interface it was executed the communication of the usage conditions of the vehicle (in this case a Nissan Van E-NV200) located in the island of Favignana and to historicize them inside the Blockchain.

6.6 Blorin Smart Contracts as Chaincodes

In this section we report few Blorin scenarios for DR and V2G.

6.6.1 Smart Contracts for Demand Response

The SC for DR is structured in the same way as the one described in Chap. 4, but implemented in NodeJS. The SC is needed to implement the DR event logic and

Fig. 6.23 OBD port in the vehicle

Fig. 6.24 OBD reader

establish the roles of the various actors on the BloRin network. In the case of DR, the actors involved will be the DSO and the users who decide to join the service by providing flexible loads. The DSO will notify the DR event on the BloRin blockchain with the aim of increasing the efficiency of the power plant or mitigating any issues expected in a given hour due to production from RES, while the users, thanks to an EMS developed for the purpose, will be able to automatically respond to the request by switching off some loads in the hours when the DR event is expected. The EMS will be embedded in the blorin ClientApp described before. Finally, the SC will issue tokens to users who have responded to the request. Two scenarios have been defined to validate DR in Blorin, described in the following.

6.6.1.1 Scenario 1

The first scenario refers to the current state of the island of Lampedusa, which produces almost all of its energy needs with a diesel power plant, consisting of 8 diesel generators whose total rated power is 22.2 MW [4]. This generation system has remained almost the same even with the presence of Photovoltaic, in 2020 about 142 kWp on the island [5]. Considering the current island state, DR could contribute to increasing the efficiency of the power plant by avoiding turning on the peak generators or avoiding the operation of the generators at low load and therefore low efficiency. The DSO can request a load reduction during hours when the peak generators are scheduled to be turned on or a load increase during hours when the power plant is scheduled to operate at low load. The BloRin blockchain aggregate users that participate in DR events by controlling water heaters, heat pumps, and washing machines, which are the most frequent loads in Lampedusa homes.

The flowchart in Fig. 6.25 shows the steps to follow for executing a DR event.

The first step is to record the daily load or production profiles of users who are part of the network on the blockchain. This step is performed automatically by the EMS embedded in the blorin ClientApp, which reads the data from the meter and through the SC sends it to the blockchain. When the DSO needs to reduce the load downstream of a substation, it notifies on the blockchain a DR event consisting of the total amount of load to be reduced and the time slot during which the reduction is required. At this point, the SC, before distributing the request to the users, evaluates the users' baselines and the users' availability to participate in the DR events using the same methods outlined in Chap. 4. According to the baseline and availability, the SC distributes the requested reduction to all users who have joined the service. The EMS reads the assigned reduction from the blockchain and based on the user's needs, it controls through smart switches or smart plugs the loads that the user has decided to make available in order to satisfy the DSO's request. After the DR event, the SC verifies if the users have correctly responded to the request and evaluates the remuneration with the same logic showed in the Chap. 4.

Fig. 6.25 DR flowchart

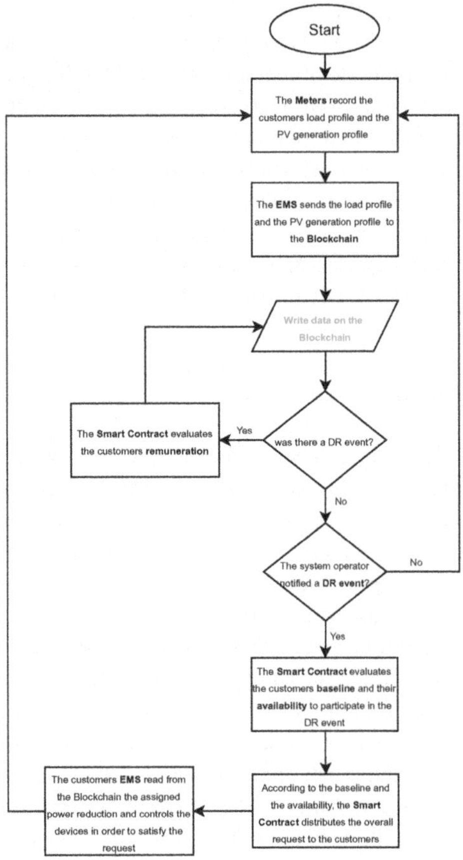

6.6.1.2 Scenario 2

The second scenario refers to the future state of the island following participation in the BloRin project. As part of the project, between 75 and 115 kWp of residential PV systems distributed downstream of four MV/LV substations feeding residential end-users will be installed by the end of 2022. Assuming maximum capacity is installed, each residential substation is considered to have an increase of 29 kWp of connected PV power. This new PV production will affect the substation load profiles, revealing possible problems, such as the appearance of steep load ramps and the risk of reverse power flows, see Fig. 6.26.

To mitigate the effects of PV power, different control logics can be implemented to control household loads such as water heaters. Such logics will be executed through, for example, the following DR events:

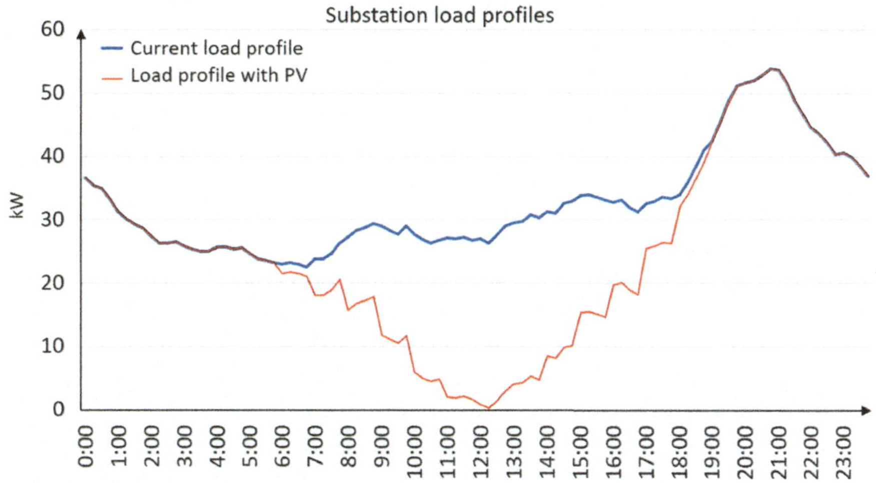

Fig. 6.26 Substation load profile shaped by PV generation

1. DR1: 50% of the water heaters are controlled to set the desired temperature to 80°C, in the time between 11 am and 6 pm. The goal is to store PV energy as thermal energy through the water heaters.
2. DR2: 35% of the water heaters are controlled to set the desired temperature to 80°C, in the period between 10 a.m. and 2 p.m., while 30% are controlled in the period between 3 p.m. and 5 p.m. During the rest of the day, the desired temperature is set to 40°C, to still provide comfort to the end users.
3. DR3: same logic as DR2, but with 50% of the water heaters controlled in the period between 10 and 18:30.

The events flow is the same as that shown in Fig. 6.25, but in this case the aim and the load management logics implemented by the EMS are different.

As can be seen from Fig. 6.27, with the control logics implemented the reverse power flow problem is avoided, while the profile shape is improved and the maximum peak is reduced.

6.6.2 Smart Contracts for V2G

The implemented scenarios exploit the studies carried out in the master thesis [6]. Following a deep analysis on the management of the daily electric load, it was possible to reduce the degree of congestion at a recharge station during the day and to offer the possibility of recharge allowing to minimize the costs (both in terms of time and money) in executing the vehicle recharge, exploiting the models elaborated in [7].

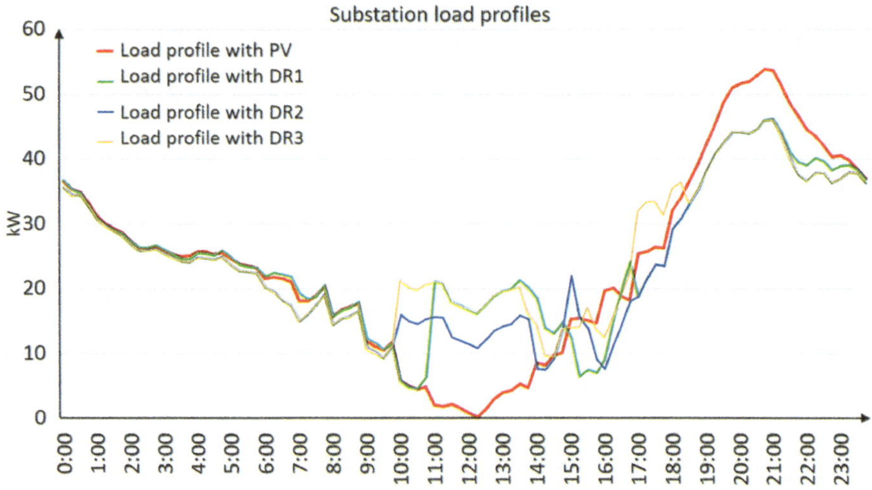

Fig. 6.27 Substation load profiles under different water heater control logics

6.6.2.1 Scenario 1

A first possible scenario includes an EV owner who is interested in having an amount of stored energy in the battery that is sufficient for a possible use of the vehicle at any time of the day. Another option considered for the EV owner was the possibility to recharge the vehicle at competitive costs. Under this latter hypothesis, thanks to a continuous monitoring of the vehicle and of the recharge stations, it is possible for the smart contract to define constraints about the number of kilometers that can still be traveled and the time for completing the recharge. It is thus decided whether to recharge the electric vehicle based on a given opportunity to recharge, starting from a given time, and to make it available within a given timeframe. This puts the driver in the condition to assess all possible options that vary along time and space according to the tariff policies in place, by the power at the recharge stations and by the distance to reach them. In the flowchart in Fig. 6.28 the applied logic is described, the algorithm regularly queries an attribute that is inside the Blockchain, given by the owner of the vehicle that sets the destination of the subsequent trip. Once the travel destination is set and the ledger is updated, it is easy to calculate whether the vehicle is sufficiently charged to cover the distance. Therefore if the algorithm recognizes that the charge is sufficient, then the algorithm enters the stand-by mode waiting for a good chance to recharge (i.e.: low cost, limited distance); otherwise, the search for a recharge station starts, reducing the costs for the electric vehicle recharge.

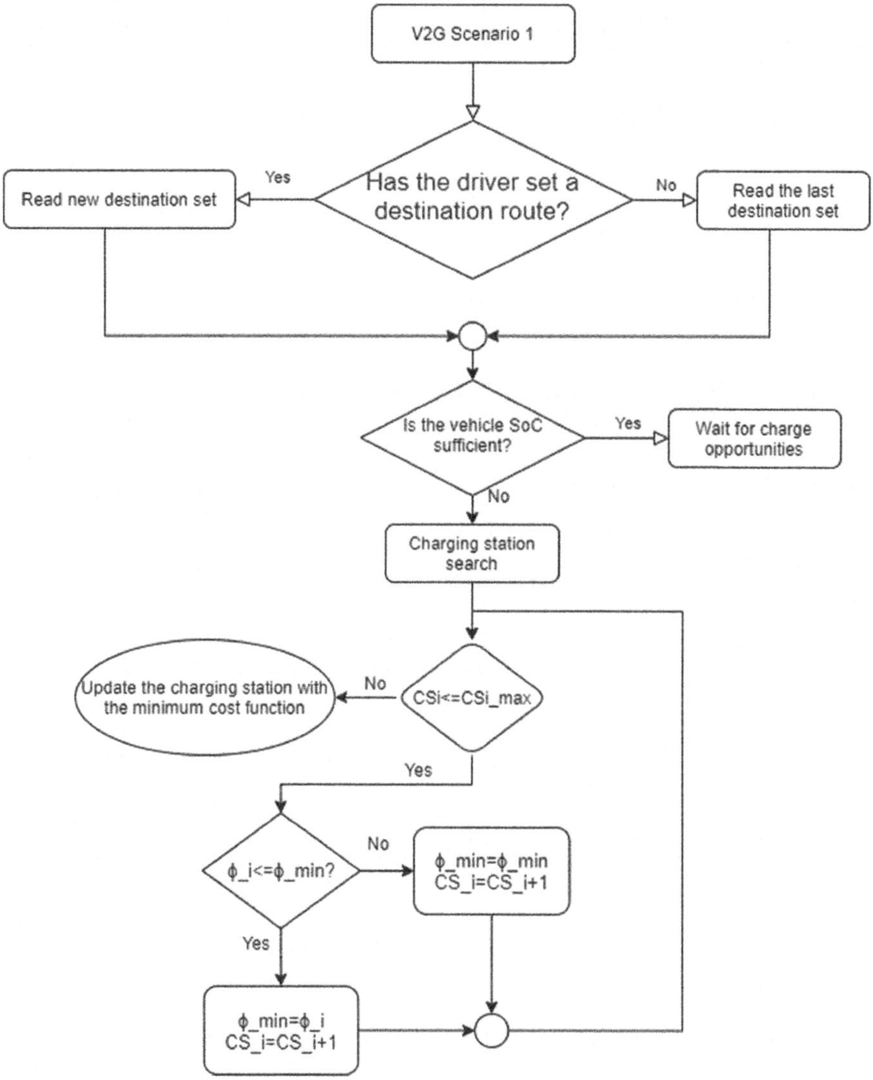

Fig. 6.28 Scenario of search of minimum cost

6.6.2.2 Scenario 2

Another perspective to be considered is the one that focuses the interests of the grid operators. In this case, their main interest is not to overload the charging stations and thus to distribute the load as much as possible homogeneously in a given geographical area and within the day. To allow this, the following scenario was considered. The latter was exploiting the increase and the reduction of the output price of energy so as to give incentives or not based on the level of congestion of devices and lines upstream the recharge units (Fig. 6.29).

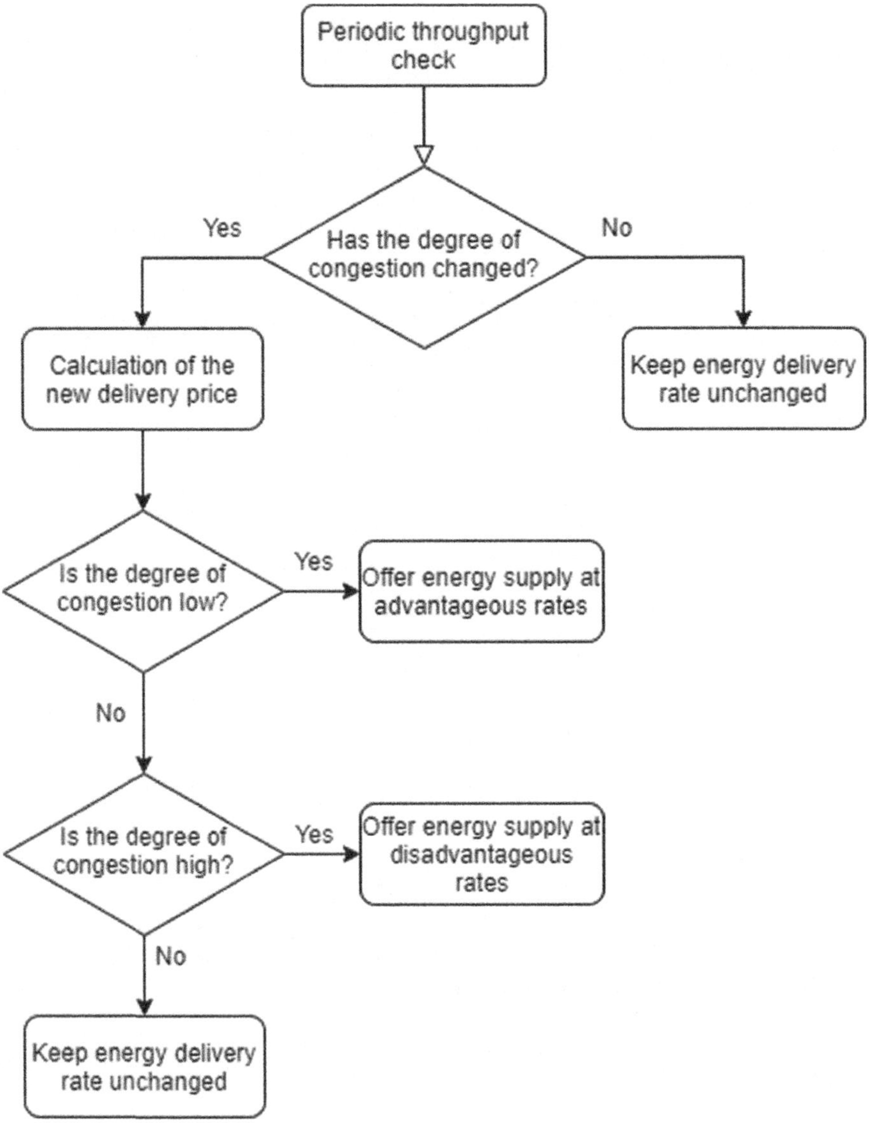

Fig. 6.29 Scenario for calculating the degree of congestion of the recharge station

6.6.2.3 Scenario 3

One of the aspects on which the focus of the project is aimed at is the possibility to allow an average user to monitor the status of health of his own electric vehicle's battery. Indeed, one of the main issues to be considered when talking about vehicle to grid services, is the problem concerning the ageing of the Lithium-Ion battery. The latter issue could be exacerbated as an effect of the use of batteries for grid services. The processes defining the ageing are small changes that are not reversible and can be classified into two main categories: gradual degradation due to ageing and that can be normally monitored and those due to a wrong usage of the vehicle and a wrong execution of the charge and discharge cycles of the electric vehicle. These bring fast changes of the battery's features. To cope with this necessity, it was decided to create two different channels that independently monitor the normal ageing of the battery and another one monitoring spatial events that influence more evidently the health status of the battery. The most significant ageing factors are listed below (Fig. 6.30):

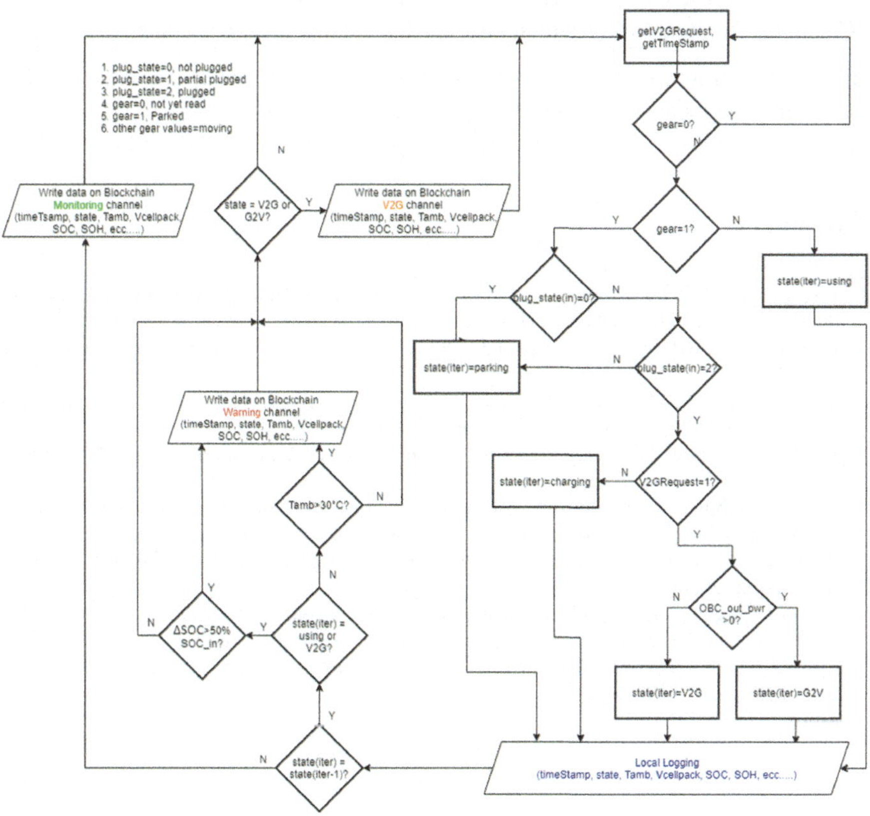

Fig. 6.30 Scenario monitoring of the SoH of the electric vehicle

- Ambient temperature influences the degradation; the higher the temperature and the higher is the degradation, especially when high temperatures are associated to high values of SoC;
- Greater the speed of charge/recharge, greater the degradation;
- The charge beyond a given level or the discharge under a given threshold (depth of discharge) can accelerate the battery degradation and reduce the lifetime.

Considering all these items, the scenario depicted in Fig. 6.30 was outlined.

Functions have thus been implemented so as to allow the update of the parameters such as the output voltage, the vehicle state, the ambient temperature, the internal temperature and the status of charge. Subsequently, exploiting the update of these parameters it was possible to execute the calculation of the variations of the state of charge and to monitor the vehicle's external temperature and record the threshold exceeding in a Warning channel. Summarizing, the considered scenario allows to:

- Monitor the electric vehicle's battery status (State of Health, SoH)
- Verify is the vehicle is in V2G or G2V statuses
- Historicize a bad use of the EV in a warning channel
- Calculate the duration for which the vehicle remains in a given state.

This allows in a future perspective to keep track of the lifetime of the battery and to connect this degradation to the lifetime events.

References

1. Empowering App Development for Developers | Docker. [Online]. Available https://www.docker.com/. Visited on 8 Feb 2021
2. Download—ZeroTier. [Online]. Available https://www.zerotier.com/download/. Visited on 8 Feb 2021
3. Docker swarm init | Docker Documentation. [Online]. Available https://docs.docker.com/engine/reference/commandline/swarm_init/. Visited on 8 Feb 2021
4. SELIS Lampedusa SPA. [Online]. Available http://www.selislampedusa.it/. Visited on 16 June 2021
5. Isole sostenibili, osservatorio sulle isole minori. [Online]. Available https://www.legambiente.it/rapporti/isole-sostenibili-osservatorio-sulle-isole-minori/. Visited on 15 June 2021
6. A. Augello, *Analisi , progettazione e sviluppo di una piattaforma blockchain per il V2G*. Ph.D. dissertation (2020)
7. P. Gallo, Charging electric vehicles using opportunistic stopovers, in *2017 IEEE International Conference on Environment and Electrical Engineering and 2017 IEEE Industrial and Commercial Power Systems Europe (EEEIC /I CPS Europe)* (2017), pp. 1–5. https://doi.org/10.1109/EEEIC.2017.7977832

The manufacturer's authorised representative in the EU is Springer Nature Customer Service Centre GmbH, Europaplatz 3, 69115 Heidelberg, Germany. If you have any concerns regarding our products, please contact ProductSafety@springernature.com

Printed and bound by CPI Group (UK) Ltd, Croydon, CR0 4YY
30/09/2026
02225262-0007